Transfer of Nuclear Technology under International Law

Transfer of Nuclear Technology under International Law

Case Study of Iraq, Iran and Israel

By

Namira Negm

MARTINUS NIJHOFF PUBLISHERS

LEIDEN • BOSTON
2009

This book is printed on acid-free paper.

Opinions expressed in the present series reflect the views of the author alone and cannot be ascribed to organizations for which the author currently works or has worked previously.

Library of Congress Cataloging-in-Publication Data

Negm, Namira.
Transfer of nuclear technology under international law : case study of Iraq, Iran and Israel / by Namira Negm.
p. cm.
Includes bibliographical references (p.) and index.
ISBN 978-90-04-17527-3 (hardback : alk. paper)
1. Nuclear nonproliferation–Middle East. 2. Nuclear-weapon-free zones–Middle East. 3. Nuclear energy–Government Policy–Israel. 4. Nuclear energy–Government Policy–Iraq. 5. Nuclear energy–Government Policy–Iran. I. Title.

KZ5770.N44 2009
341.7'340956–dc22

2009017263

ISBN 978 90 04 17527 3

PRINTED IN THE NETHERLANDS

To my parents
Nabil Negm and Nadia Diab

CONTENTS

ACRONYMS AND ABBREVIATIONS

AEC	Atomic Energy Commission (US)
AEOI	Atomic Energy Organisation of Iran (Iran)
AFRA	African Regional Cooperative Agreement for Research, Development and Training Related to Nuclear Science and Technology
AG	Aktiengesellschaft (corporation)
BAW	Federal Economics Office (Germany)
BG	Board of Governors of IAEA
CAS	Committee on Assurances of Supply
CoB of NAM	Coordination Bureau of the Non-Aligned Movement in New York
CD	Conference on Disarmament
CERN	European Organisation for Nuclear Research
CIA	Central Intelligence Agency (US)
CNC	Computer Numerically Controlled machines
CTBT	Comprehensive Test Ban Treaty
DG	Director General of the IAEA
DOE	Department of Energy (US)
ECOSOC	Economic and Social Council
EDC	Emerging Design Centre (Iran)
EMIS	Electromagnetic Isotope Separation
ENACE	Empresa Nuclear Argentina de Centrales Electricas (Argentina's Nuclear Company for Power Plants)
ENTC	Esfahan Nuclear Technology Centre (Iran)
ERDA	Energy Research and Development Administration (US)
EU	European Union
EURATOM	European Atomic Energy Agency (European Union)
FRG	Federal Republic of Germany (West Germany)
GA	General Assembly
GC	General Conference of the IAEA
GHG	Greenhouse Gas
GIE	Groupement d'Interet Economique (Group of Economic Interests)

GMBH	Gesellschaft mit beschränkter Haftung (company with limited liability)
GOV	Governmental
GPS	Global Positioning System
HEU	Highly Enriched Uranium
HTGR	High Temperature Gas-cooled Reactor
HWR	Heavy Water Reactor
IAEA	International Atomic Energy Agency
IAEA	NTR Nuclear Technology Review
IAEC	Israel Atomic Energy Commission (Israel)
ICJ	International Court of Justice
ICR	Inventory Change Report
ICTY	International Criminal Tribunal for the Former Yugoslavia
INFCIRC	Information Circular
INVO	IAEA's Iraq Nuclear Verification Office
IPC	Industrial Projects Company (Iraq)
IRT 2000	A type of research reactor
JHL	Jabr Ibn Hayan Multi Purpose Laboratory (Iran)
KWH	(e) Kilo Watt/ Hour- Electric
KWU	Kraftwerk Union (Germany)
LEU	Low Enriched Uranium
LIS	laser isotope separation
LMFR	Liquid Metal- cooled Reactor
LOF	locations outside of facilities
LWR	Light Water Reactor
MIX	Molybdenum, Iodine and Xenon Radioisotope Production Facility (Iran)
MW	Megawatt
Mwe	Megawatt- Electric
NEA	Nuclear Energy Agency (OECD)
NEA	*the Regulation of Nuclear Trade*, 1988
NNPA	Non-Nuclear Proliferation Act (US)
NPT	The Treaty on the Non-Proliferation of Nuclear Weapons
NPTRC	Treaty on the Non-Proliferation of Nuclear Weapons Review Conference
NPR	Nuclear Posture Review
NRC	Nuclear Regulatory Commission (US)
NUMEC	Nuclear Material and Equipment Corporation (US)
NWFZ	Nuclear Weapon Free Zone

NNWS	Non-Nuclear Weapon State(s)
NSG	Nuclear Supplier Group (London Club)
NWS	Nuclear Weapon State(s)
OECD	Organisation for Economic Cooperation and Development (Europe)
OMV	Ongoing Monitoring and Verification System on Iraq
OPEC	Organisation of the Petroleum Exporting Countries
OPEN	Organisation des Producteurs d'Energies Nucléaire (Organisation of Nuclear Energy Producers)
PC3	Petrochemical 3 (Iraq)
P5	Permanent Member States to the Security Council
PUNE	Peaceful Uses of Nuclear Energy Conference
R&D	Research and Development
RCA	Regional Co-operative Agreement for the Advancement of Nuclear Science and Technology for Asia and the Pacific
SG	Secretary-General of the United Nations
SC	Security Council
SCR	Security Council Resolution(s)
SIPRI	Stockholm International Peace Research Institute
SWU	Separative Work Units
TDG	Technology Development Group (UK)
TNRC	Tehran Nuclear Research Centre (Iran)
TRR	Tehran Research Reactor (Iran)
U-233	Uranium-233
U-235	Uranium-235
U-238	Uranium-238
UF	Uranium Floride
UF6	Uranium Hexafloride
UK	United Kingdom
UN	United Nations
UN 1994	Assuring the Success of the Non-Proliferation Treaty Extension Conference, 1994
UNSCOM	United Nations Special Commission on Iraq (Iraq)
UNIDIR	United Nations Institute for Disarmament Research
UNMOVIC	United Nations Monitoring, Verification and Inspection Committee
UO	Uranium Oxide
US	United States of America
USSR	Union of Soviet Social Republic

VAT	Vakuumventile AG
WEA	World Energy Assessment
WMD	Weapons of Mass Destruction
WHO	World Health Organisation

INTRODUCTION

To be, or not to be: that is the question. William Shakespeare

International law is more than just an ephemeral debating point. In disarmament, as in other salient aspects of international life, law has meaning and impact. It guides behaviours, shapes expectations, undergirds notions of appropriateness and legitimacy, and, in the extreme, provides at least an imperfect collective format for redress of violations and other grievances. In the area of arms non-proliferation, in particular, law can be one of the sinews that bind the community of nations, contributing valuably to the overall effort to retard the spread of dangerous weaponry. David A. Koplow[1]

The "Atoms for Peace" project that President Eisenhower launched in 1953, the establishment of the IAEA in 1957 and the adoption of the NPT in 1968 were all directed towards encouraging the transfer of nuclear technology for peaceful purposes and simultaneously headed for prohibiting the spread of knowledge regarding the military uses of the atom.

This book focuses on assessing the rules and implementation of the NPT, the IAEA, the arrangements of the Supplier Groups and international cooperation to transfer nuclear technology. It seeks to answer the subsequent questions:

- Are the rules of the NPT and IAEA adequate enough to guarantee the transfer of nuclear technology for peaceful purposes?
- Does the role of the Supplier Groups affect negatively the transfer of nuclear technology for peaceful purposes?
- Are the IAEA measures to tighten its safeguards sufficient to prevent the diversion of the transferred material from peaceful to military uses?

[1] Koplow, David A., 'The Jurisprudence of Non-Proliferation: Taking International Law Seriously,' Transnational Law and Contemporary Problems, Volume 2, No. 2, Fall 1992, p. 358.

- What are the consequences of the indefinite extension of the NPT?
- How does the delay in implementing Article VI of the NPT affect the non-universality of the Treaty?
- Are there inherent failures in international rules that govern non-proliferation of nuclear weapons?
- If these failures exist, do they affect the conduct of States in the course of implementing these rules?
- With regards to the Middle East, do the failures in the rules or its implementation affect the establishment of an NWFZ in the region?

The study shall investigate how the inadequacies in the international legal system seem to be among other reasons behind the failure to establish an NWFZ in the Middle East. Israel, the only NWS in the region, claims that the system fails to provide sufficient security assurances. A nuclear Israel triggered the militarisation of the Iraqi nuclear programme. Such was a development that the NPT/IAEA safeguards failed to timely detect. Iran is another case where the system could not preserve its inalienable right as an NNWS to access peaceful nuclear technology.

The book shall also assess if any amendments are required to the NPT rules in order to bridge up the gaps created by its implementation, and if so, what is the effect of such amendments on establishing an NWFZ in the Middle East.

The current study focuses on discussing the previous issues as follows:

The First Chapter highlights the importance of the peaceful uses of nuclear technology, not only for generating electricity, but also for using nuclear techniques for the benefit of humankind. These techniques comprise applications in the fields of medicine, agriculture, irrigation, industries, crime detection, food production, insect and pest control, increase of animal reproduction and more. Nuclear knowledge, as such, enhances scientific progress in a country as it creates an assemblage of scientific elite that are involved in different aspects of life.

Nuclear technology has positive effects on human lives. However, it is not accessible or affordable to all States. For some States, electricity generated from nuclear power is not cheap. Developing States have access to nuclear techniques through the Technical Cooperation Programme of the IAEA and often through bilateral cooperation with developed States. Unfortunately, nuclear research is limited in developing States as compared to developed ones.

Another factor affecting the use of nuclear technology lies in its lack of public acceptance. The involvement of such technology in our lives is still limited because of several public concerns that have not been fully addressed, particularly waste problems as well as safety and environmental effects.

The Second Chapter deals with the rules provided by international legal instruments to encourage the transfer of nuclear technology for peaceful purposes. It is divided into three sections as follows:

The First Section assesses the rules of the NPT concerning the transfer of nuclear technology for peaceful purposes and the disarmament obligations the Treaty imposes on both NWS and NNWS. The NPT's main objective is the entire elimination of nuclear weapons. It conditions the transfer of nuclear technology upon its non-diversion from peaceful to military purposes. In addition, it requests NNWS to negotiate Safeguards Agreements with the IAEA to ensure that any nuclear activity they undertake remains solely devoted to peaceful purposes.

Article IV of the Treaty addresses the inalienable right of its parties to develop research, production and use of nuclear energy for peaceful purposes without discrimination. It also requests them to facilitate and participate in, the fullest possible exchange of equipment, materials as well as scientific and technological information for the peaceful uses of nuclear energy. Moreover, it encourages the contributing States, party to the Treaty, to give due consideration to the needs of developing areas of the world, particularly NNWS party to the Treaty, while contributing alone or together with other States or international organisations to promote the transfer of nuclear energy for peaceful purposes.

Article V requests NWS to share the potential benefits of peaceful applications of nuclear explosions with NNWS party to the Treaty on a non-discriminatory basis. However, due to the conclusion of a Comprehensive Nuclear Test Ban Treaty in 1996 and the cessation of nuclear tests, Article V is no longer put into practice.

The disarmament obligation of NNWS is a passive one. They are obligated not to divert nuclear energy from peaceful uses to nuclear weapons or to any other nuclear explosive devices (Article IV). Furthermore, NNWS are required to refrain from receiving, gaining control of, manufacturing, acquiring, seeking or receiving assistance to manufacture nuclear weapons or other nuclear explosive devices. (Article II).

As for NWS obligations, the first is not to assist, in any manner, an NNWS in manufacturing or receiving a nuclear explosive device and not to provide materials and equipment that may assist in producing nuclear weapons (Article III). The second obligation is provided by Article VI which affirms the purpose of the Treaty as stipulated in the preamble, whereby the parties to the Treaty declare 'their intention to achieve, at the earliest possible date the cessation of the nuclear arms race and to undertake effective measures in the direction of nuclear disarmament.'[2]

These obligations, together with the controversial and discriminatory character of the NPT, are assessed in details. The Treaty is a compromise between NWS and NNWS. As previously mentioned, NNWS agree to restrict their activities to the development of nuclear capabilities for peaceful purposes, in return for an obligation imposed, in Article VI of the Treaty, on the NWS to pursue negotiations in good faith to end the nuclear arms race and to eliminate nuclear weapons. However, the NPT did not propose any concrete measures to enforce this obligation.

The book provides an assessment of the effects of the indefinite extension of the NPT, which so far has not eliminated threats to the system but instead has merely insured its flawed survival.

The reasons behind the lack of universality are also discussed. Several States continue to act in defiance of the current system and to operate outside it. These States are India, Pakistan, Israel and the case of North Korea. Achieving universality will signify a major step towards the complete elimination of nuclear weapons.

The Second Section deals with IAEA, which was established in 1957 to ensure that nuclear technology is not employed to further any military use. According to Article II of the IAEA Statute; the Agency has a dual-objective role: it shall seek to accelerate and enlarge the contribution of atomic energy to peace, health and prosperity throughout the world; and it shall ensure, so far as it is able to, that the assistance provided by it or at its request, under its supervision or control, is not used in such a way as to further any military purpose.

The First Chapter of the study examines the fields of cooperation between the IAEA and its member States which are directed towards improving the latter's scientific, technological and regulatory capabilities through effective implementation of the IAEA's Technical Cooperation Programme.

The other role of the Agency, to ensure that no nuclear material or equipment transferred to an NNWS will be diverted and used in manufacturing a nuclear weapon or any other nuclear explosive device under the IAEA Safeguards, is detailed. The NPT safeguards system grants the IAEA authority to conduct investigative activities that include gathering

---

[2] Article VI of the NPT.

information and inspecting in NNWS to verify the accuracy of the States' declarations. It also obligates NNWS to cooperate with the Agency in facilitating its work in their territories.

The two IAEA safeguards systems will be scrutinised. Before the NPT, safeguards were implemented on case by case basis upon the request of either the supplying state or the recipient state. This system is still applied to the nuclear facilities of non-NPT members. The rights and obligations of the Agency under Safeguards Agreements (INFCIRC/66), concluded with States before the NPT, will be discussed.

The second type of Safeguards systems is that of the NPT. The NNWS members to the NPT are required to conclude safeguards agreements with the IAEA (under the model NPT Safeguards Agreement INFCIRC/153). According to Article III of the NPT, NNWS undertake to accept IAEA safeguards "on all source or special fissionable material in all peaceful nuclear activities within its territory, under its jurisdiction or carried out under its control anywhere, for the exclusive purpose of verifying that such material is not diverted to nuclear weapons or other nuclear explosive devices."

INFCIRC/153 illustrates the main inspection rights and obligations of the IAEA. Under the Agreement, an NNWS has to report to the Agency its nuclear activities and the Agency has the right to conduct several types of inspections to verify that nuclear activities in the member States remain peaceful: regular, ad hoc and special inspections.

Constraints on the Agency's legal authority to gain access to information and sites will be discussed, especially the balance between IAEA rights and State's sovereignty.

Development in the Agency's inspections techniques were initiated by the revelation of the Iraqi clandestine nuclear programme. The new techniques should assist the Agency in the timely detection of any diversion of nuclear material in an NNWS from peaceful to military uses. There were no legal problems to implement a number of these techniques under INFCIRC/153. However, certain other techniques needed a different legal frame for their implementation. Hence, in 1993 the IAEA Secretariat prepared a comprehensive plan called "Programme 93 + 2". The plan, which was endorsed by the Board of Governors of the IAEA (BG) in 1995, called for more inspection rights, including more access to information and sites. It came out in 1997 in the form of the Model Additional Protocol to Safeguards Agreements (INFCIRC/540).

The adequacy of the safeguards rules to ensure the non-emergence of new NWS in the future will be assessed, along with the new developments in their implementation.

The Third Section discusses the transfer of nuclear technology for peaceful purposes through regional and bilateral cooperation. Bilateral cooperation represents the main channel for providing nuclear technology around the world. International regional cooperation plays an important role in exchanging scientific knowledge, research and development in the nuclear field. Some of the principle avenues for international cooperation are organisations, such as the Nuclear Energy Agency (NEA) and European Organisation for Nuclear Research (CERN). European Atomic Energy Agency (EURATOM) is another agency with political objectives that focuses on encouraging the establishment and growth of nuclear industries in the territories of its member States.

This Section also discusses the effects of the national regulations of the US, as the major supplier of nuclear technology around the world, on international cooperation in the field. It assesses the approach undertaken by a Supplier State, while disregarding its national legal rules to serve different political interests.

Chapter 3 studies the guidelines of the Supplier Groups and their effects on the transfer of nuclear technology to NNWS. They work through gentlemen agreements to reach their goals. Supplier Groups claim that they assist the IAEA to ensure the non-proliferation of nuclear weapons. However, NNWS regard these groups as cartels established to maintain the monopoly of nuclear suppliers around the world.

The major Supplier Groups are the Zangger Committee and the Nuclear Supplier Group (NSG). Their guidelines are a set of identical but unilaterally adopted voluntary undertakings that implement certain conditions restraining the transfer or re-transfer of sensitive nuclear technology, equipment, and materials. These conditions include assurances of non-explosive use by the recipient, effective physical protection for nuclear materials and IAEA safeguards agreements, as well as other measures also intended to limit the spread of nuclear weapons.

The Supplier Groups, in order to agree to a supply request, study the safeguards status in the requesting State. They do not merely rely on its membership of the NPT or its conclusion of safeguards agreements with the IAEA. Instead, they go beyond these requirements by assessing the implementation of safeguards in the requesting State and evaluating the sufficiency of inspections to ensure the non-diversion of nuclear technology into military projects. This individual evaluation uses subjective criteria, which in turn affect nuclear supplies. The book examines such effects.

The book also discusses the history of nuclear trade in brief, the emergence of new nuclear suppliers highlighting the effects on nuclear trade.

Although the traditional suppliers are more developed and more credible in terms of safety and technical capabilities, the emerging suppliers seem to be more appealing to developing countries because they provide cheaper material, equipment and know-how.

Emerging suppliers are believed to be a threat to the non-proliferation regime. However, most of them are now members of the NSG and apply the same guidelines as traditional suppliers. In the end, all suppliers can be regarded as proliferators from one perspective, and as providers of nuclear knowledge and technology from the other.

Within the discussion of the previous issues, the study will cover the failures and inadequacies in the NPT/IAEA rules. These failures include the inherent contradiction between law and State sovereignty, similar to the contradiction between the law and individual personal freedoms. The law as a social science restricts the behaviour of States. In this context, the contradiction between IAEA rights and State sovereignty and the efforts of the IAEA to strike a balance between them will be assessed.

The failure to strike this balance sometimes leads to discrepancies in cooperation between States and the IAEA. How and where these discrepancies and others exist will be assessed to determine the effects on the case studies and on the establishment of an NWFZ in the Middle East.

Chapter 4 summarises the Israeli nuclear military programme. It assesses the legal obligations of Israel under the relevant SC and GA resolutions.

Israel is not a party to the NPT and its nuclear facilities do not fall under IAEA safeguards, except in limited features under INFCIRC/66 (the Model IAEA safeguards agreement adopted before the conclusion of the NPT, and designed to be implemented upon the request of either a supplier or a recipient State).

Israel formally follows a policy of nuclear ambiguity; the Israeli Government maintains the position that it "will not be the first to introduce nuclear weapons into the Middle East region."[3]

Nevertheless, the account of an Israeli technician who worked at the Dimona Reactor in the 1980's and the statement delivered by the Israeli Prime Minister Ehud Olmert in a TV programme in which he listed Israel

[3] Cohen, Avner, 'Beyond the Pale: Israel and the Bomb', Bulletin of Atomic Scientists, Volume 55, Issue 1, January/ February 1999. He stated that Lev Eshkol, the third Prime Minister of Israel, was the author of this statement.

among NWS, confirmed the suspicions of the international community that Israel manufactured nuclear weapons, but it continues to follow the "bomb in the basement" doctrine.

Israel claims that it has legal, political and security reasons to preserve its nuclear weapons as a deterrent against its adversaries and to maintain its position as a non-NPT member. Several NPT failures and inadequacies of IAEA safeguards are among the reasons Israel points to while defying the regime.

A nuclear armed Israel created an imbalance of power in the Middle East, which was subsequently among the forces driving other States, like Iraq and Libya, to seek to develop nuclear capabilities for military purposes. Israel is also keen to preserve its nuclear monopoly in the region, and therefore supports efforts to slow down the flow of nuclear technology into the region for peaceful purposes, as with the case of Iran.

Chapter 5 discusses the Iraqi nuclear programme. Although a member of the NPT and of the IAEA, Iraq sought to acquire nuclear technology to establish sufficient infrastructure to produce a nuclear bomb.

The discovery of the Iraqi nuclear programme came as a shock to the international community. The surprise was that Iraq could develop a clandestine nuclear weapon programme while its nuclear facilities had been placed under the IAEA safeguards and were inspected regularly.

The Iraqi case was a turning point in the IAEA safeguards system. The international community realised that the current IAEA/NPT safeguards system may not be sufficient to ensure that an NNWS is complying with its international obligations under the current rules.

The IAEA developed new inspection techniques and adopted the Model Additional Protocol to the Safeguards Agreement (INFCIRC/540) to impose more legal obligations on NNWS and provides IAEA with more rights to access information and sites in the territory of the member States.

In this Section, special emphasis is placed on the role of nuclear suppliers in assisting Iraq to build up its programme and yet remained unnoticed by the IAEA.

Chapter 6 indicates that the Iranian programme, according to the findings of the IAEA, is directed to peaceful uses of the atom. However, several States claim that Iran is developing nuclear weapons covertly. Nevertheless, there is no evidence, so far, that Iran is developing its nuclear capabilities to further any military purpose.

As a member of the NPT and IAEA, Iran exercises its right to acquire nuclear technology for peaceful purposes to NNWS.

The Iranian programme started as early as 1970. It was directed towards generating electricity from nuclear power reactors.

Iran's nuclear facilities are placed under IAEA/NPT safeguards. However, the Iranian Government has failed to declare to the IAEA some facilities and to report its acquisition of certain sensitive material and equipment. These failures constitute violations to Iranian obligations under international law.

Iran has undertaken several corrective measures. It is cooperating with the Agency to clarify controversial aspects of the Iranian nuclear programme. It has signed the Model Additional Protocol (INFCIR/540) and has voluntarily declared the suspension of some of its nuclear activities in respect to the enrichment of uranium activities as a confidence building measure.

The book assesses the Iranian nuclear programme, the legal obligations Iran failed to satisfy and the recent developments undertaken by both Iran and the IAEA to ensure the Iranian compliance to its legal obligations under its Safeguards Agreement.

However, the Iranian case raises several concerns related to safeguards, confidence building measures, approaches by supplier States towards an NNWS and whether these concerns may hamper the inalienable right of an NNWS to acquire nuclear technology and to build nuclear infrastructure for peaceful purposes.

My interest in the current affairs in the Middle East region was behind this study. While realising the benefits of peaceful uses of nuclear technology, and simultaneously, witnessing daily events of tensions, conflicts, threats of war and calls for disarmament, my attention was directed towards the issue of disarmament in the area, especially removing Weapons of Mass Destruction (WMD) and more specific, eliminating nuclear weapons from the region.

In assessing the current international rules governing the transfer of nuclear technology for peaceful purposes and the spread of nuclear weapons around the world, it was necessary to conduct a legal study on establishing an NWFZ in the Middle East.

However, a problem which faced the undertaking of this study was the lack of sufficient literature on the subject. Much has been written on nuclear disarmament, but little has covered the legal rules controlling the transfer of nuclear technology for peaceful purposes.

Hence, it was important to rely on primary sources like UN documents (resolutions & reports) and IAEA reports, statements and resolutions. No statistics were found regarding certain elements of the book. The several gaps found during the course of the study are due to the secrecy surrounding nuclear programmes and lack of transparency in the work of nuclear Supplier Groups. To bridge up these loopholes, several interviews with experts in the field were conducted in Belgium, Cairo, Vienna and the UN.

This study constitutes a part of the ongoing debate on the transfer of nuclear technology for peaceful purposes and the adequacy of current rules to prevent the misuse of this technology.

The revelation of a nuclear military programme and claims that Iraq was still developing WMD led to a war and a new invasion in the Middle East. If tension accelerates in the region, there are fears that other wars may break out generating a significant threat to the stability of the area. Though all previous wars in the Middle East had been strictly conventional, the introduction of WMD, especially nuclear weapons, increases threats of a non-conventional war where nuclear, chemical or biological weapons may be deployed and utilised.

This highlights the importance of the initiative to establish an NWFZ in the Middle East, originally put forward in 1974 by Iran and supported by Egypt. Both Israel and the rest of States in the region, including Arab States, support the establishment of an NWFZ in the Middle East; however, they demonstrate different approaches to the matter. Israel demands negotiations to establish an NWFZ as well as adopting fundamental steps towards peace prior to any negotiations on WMD control.

Chapter 7 analyses the two aforementioned approaches to establish an NWFZ in the Middle East. The obligations of States are assessed according to GA resolutions and the final documents of the NPT Review Conferences relevant to the implementation of this initiative. These obligations include the following:

- States Parties to the NWFZ undertake to use nuclear energy for exclusively peaceful purposes and, in order to ensure compliance with this undertaking, set up a system for controlling nuclear activities in the region.
- NWS are required not to introduce nuclear weapons into the zones and not to use or threaten to use such weapons against the zone parties through "positive" and "negative guarantees".

These obligations cannot be maintained in the Middle East unless Israel is ready to formally declare its acquisition of nuclear weapons, as a step towards renouncing them through effective measures, and to place its nuclear facilities under IAEA safeguards.

The current political situation in the Middle East delays the efforts towards establishing an NWFZ. However, if the NPT rules are amended to provide Israel with adequate security assurances and constructive measures are undertaken towards building peace in the region, the establishment of an NWFZ in the Middle East may eventually become feasible.

Hopefully, the image is not as dark as it seems and as long as there is a will, there must be a way to survive.

CHAPTER ONE

BENEFITS AND COSTS OF USING NUCLEAR TECHNOLOGY FOR PEACEFUL PURPOSES

1. *Introduction*

"Good atoms and bad atoms" is a captivating expression which is self-contradictory as well as self-explanatory. It describes the status of atoms in human minds. Upon the utterance of the word 'nuclear', a fearful mood arises. It reminds people of the horrors ensuing from the military uses of nuclear materials; meanwhile the peaceful uses of nuclear energy are rarely focused upon. Consequently, bad atoms were the ones gaining attention whereas the good atoms were marginalised.

It is essential to set out the elementary scientific facts concerning atomic fission. The nucleus of the atom consists of neutrons and protons held together by immensely strong internal forces, normally incapable of being distributed by any force smaller than that provided by bombarding it with another sub- atomic particle. When neutrons, as they exists in nature, hit the nucleus of an atom of one of the components of the heavy metal uranium, it splits into two (atomic fission). This split releases an enormous amount of energy, most of which causes the two fragments to move apart at high speed among the uranium atoms. The metal is so dense that their movement is quickly slowed down, transforming their energy of motion into heat. In part, this heat that is used in nuclear electricity stations to provide steam for the turbo-generators. In atomic fission, however, neutrons are released as well as energy, and some of these neutrons split neighbouring nuclei, spreading the fission reaction through mass uranium. For the nuclei to split most effectively, the bombardment must be by low-energy neutrons, whereas those released by fission are of very high energy. They are therefore slowed down, between their release and their impact with neighbouring fuel, by arranging for them to collide with the atoms of a suitable 'moderating' material such as graphite, until their thermal energy has been sufficiently reduced. This is in practice achieved by arranging the pellets or rods of uranium fuel appropriately spaced in a block of moderator. Natural uranium is made up of two sorts (isotopes) of atoms, uranium-235 (U-235) which can be split by low-energy neutrons, and uranium-238 (U-238) which virtually cannot be split. Only one atom of U-235 exists in 140 atoms of natural uranium. It has been found that if some of the liberated neutrons enter atoms of the non-fissile U-238, they become atoms of plutonium, which is the fissile; thereby the possibility to utilise a larger proportion of the atomic energy in natural uranium. There is also a third isotope, uranium-233 (U-233), which does not exist naturally but is produced by bombarding thorium with neutrons, and is also fissile. There are thus three known nuclear fuels, U-235 that occurs in nature, plutonium, and U-233 that can be produced by neutron bombardment of thorium in nuclear reactors.[1]

In this Chapter, the role of good atoms in human life will be discussed, including the use of nuclear energy as a source of power to generate electricity, in industry, agriculture, water management, medicine, animal breeding, pesticides, crime detection and other peaceful uses of nuclear technology for the benefit of humankind. A special emphasis will be given to the costs of acquiring nuclear technology in order to reach a better understanding of the advantages and disadvantages of developing and using nuclear technology and to evaluate the costs that nations pay to obtain it.

Nuclear technology can be used to promote scientific progress and has a positive effect on the economic prosperity of nations that utilise atomic energy for their socio-economic welfare. Atomic energy is capable of producing electricity at a high economic level and has yet to be used on a large scale. In a survey of key technologies for the 21st century, physicist Freeman Dyson saw "the age of nuclear power as a symbol of exaggerated expectations and broken promises".[2] This statement makes sense in the case of our recognition that after decades of commercial use of nuclear power, it constitutes only 17 % of world electricity.[3] This renders several

[1] Plant, Sir Arnold 'The Economics of Nuclear Energy' published in 'Social implications of the Peaceful Uses of Nuclear Energy,' 1972, pp. 130–131.

[2] Young, Warren, 'Atomic Energy Costing,' 1998, p. 4 quoting from Dyson Freeman, Scientific American 150th Anniversary Issue: 'Key technologies for the 21st Century', 1995, p. 88.

[3] Ibid. Also the NPT Review Conference, 'Final Document of the 2000 Review Conference of the Parties to the NPT,' NPT/CONF.2000/28, 24 May 2000, stated the same figure in p. 4, "Nuclear power is an important contributor to the world's electricity needs. In 1999, it supplied more than one sixth of global electricity and a substantial 31 % of the electricity in Western Europe."

considerations: the excessive demand of power in industry, the development of generating electricity from nuclear power, its costs and its social implications.

Nuclear techniques can be and are used for digging canals and for irrigation purposes. It helps agriculture, produces insecticides, and is very useful in several industrial processes also. In the treatment of cancer, the use of atomic energy has yielded good results. In addition, nuclear energy is used in crime detection with success.[4]

In the manufacture of certain metals, like aluminium, and in industries involving chlorine, phosphates, fertilisers, cement, glass, iron and steel, brick-making and locomotive energy, the use of atomic energy is highly profitable.[5]

It can be used in augmenting water-supply in rivers and thus helping irrigation and food production. A river, which is dry in winter, can be supplied with water by arranging a suitable input of heat in the snow reserves in the catchment areas of the river. Then the fission device would obviate the necessity of building reservoirs for storing water.[6]

The Final Document of the 2000 Review Conference recognised the benefits of the peaceful applications of nuclear power and nuclear techniques illustrating their contribution in achieving sustainable development in developing countries and for generally improving the well-being and the quality of life of the peoples of the world.[7]

2. *Peaceful Uses of Nuclear Technology (Application of Nuclear Techniques and Nuclear Power)*

In this section, emphasis will be given to some of the uses of nuclear techniques in the fields of animal reproduction, insect and pest control, quality and safety of food, soil and water management and crop nutrition, plant breeding and genetics as well as climate change issues. These techniques are implemented through the IAEA technical cooperation programs.

---

4 Sethna, J., 'International Legal Controls and Sanctions Concerning the Production and Use of Atomic Energy,' 1966, p. 120.

5 Ibid, pp. 120–121.

6 Ibid, p. 121.

7 NPT Review Conference, 'Final Document of the 2000 Review Conference of the Parties to the NPT,' NPT/CONF.2000/28, 24 May 2000, p. 14.

2.1. *Use of Nuclear Techniques to Increase Animal Reproduction*[8]

The productivity of livestock in developing countries is often much lower than in developed countries. Livestock production in developing countries can be improved if attention is given to optimising animal nutrition and enhancing reproductive performance and health. Nuclear and related techniques can be used as tools to monitor these parameters and to measure the effects of bio-technology to improve livestock productivity.

Nuclear and related techniques can be used as tools in nutrition studies to help understand: how food materials are digested, how different nutrients are utilised, and what nutrients are deficient or unbalanced.

They can be used in studies of reproductive performance to determine the onset of puberty, the resumption of ovarian functions after birth, ovarian dysfunctions, the efficiency of oestrus detection and non-pregnancy following breeding.

Furthermore, they can be used as diagnostic aids for major animal diseases such as foot and mouth disease, contagious bovine pleuropneumonia, Newcastle disease and Rinderspest (the latter was fought successfully in Africa).

Nuclear and related bio-technologies are used to identify constraints on productivity and to find sustainable solutions, as well as to facilitate the development and dissemination of technologies, guidelines and standards which enhance food security through sustainable improvements in productivity and income generation from domestic and international trade in livestock and livestock products.[9]

The main emphasis is to improve nutrition, reproductive efficiency and health of ruminants and poultry kept on mixed farming and peri-urban based systems through a more efficient natural and human resources utilisation.[10]

2.2. *Insect and Pest Control*

Insect pests have devastating effects on crop production and on transferring diseases that affect crops and livestock. The conservative estimates

---

[8] IAEA, 'FAO/IAEA Training and Reference Centre—General Information.'

[9] IAEA, 'Joint FAO/IAEA Programme, Nuclear Techniques in Food and Agriculture—Supporting Research for Agricultural Development,' Sub-Programme on 'Animal Production and Health.'

[10] Ibid.

indicate that pests reduce world food production by 25% to 30%. Concerns over the environment and insecticides resistance have grown considerably. In this regard, in order to reduce insecticide use and its negative impact on the environment, the development and implementation of environmentally-friendly nuclear techniques for integrated area-wide management of insect pests should be processed and used.[11]

The nuclear technique is applied in this field through the use of sterile insect technique in which large numbers of insects are reared in large bio-factories sterilised by ionising radiation and released into a wild population. The wild females are sterilised following mating with a released sterile male leading to population collapse and eventual extinction.[12]

This technique succeeded in fighting two agriculture pests of major importance: the Medfly and Tsetse. The first is a pest of fruit and vegetables in sub- tropical countries and a key quarantine pest for the movement of agricultural products. While the latter is a vector of trypanosomes (parasite microscopic unicellular animals causing sleepiness disease) which causes sleeping sickness in humans and Nagana (a tropical disease of cattle transmitted by Tsetse) in cattle. They infest 10 km^2 in sub Saharan Africa and are a major cause of agricultural stagnation in the continent. This technique was used against other key pests like Screw-worm, Tephritid fruit flies and major Lepidopteran pests. It succeeded in fighting these pests through systems of mass-rearing, genetic sexing, quality control, aerial release and field monitoring for more cost-effective implementation of these techniques against key insect pests.[13]

Nuclear techniques can be used in controlling or eradicating major insect pests of crops, as well as insects of veterinary importance. Henceforth generating a reduction in pesticide use and their negative impacts on food and environment, to preserve biological diversity and contribute to the introduction of profitable and sustainable agricultural systems in developing countries.[14]

---

[11] IAEA, 'Joint FAO/IAEA Programme, Nuclear Techniques in Food and Agriculture—Supporting Research for Agricultural Development,' Sub Programme on 'Insect and Pest Control.'

[12] Ibid.

[13] Ibid.

[14] Ibid.

2.3. *Quality and Safety of Food*

The provisions of the agreement on the application of Sanitary and Phytosanitary Measures require states to monitor the potential microbiological and chemical food contamination as well as the necessity to monitor the quality of pesticide products in trade and the level of pesticide residues in food and in environment. This task can be achieved by applying nuclear and related bio-technological analytical method to obtain sustainable food security by improving food availability, quality and safety.[15]

Nuclear techniques are applied here through using irradiation as a method to reduce post harvest food losses, to enhance food safety, and to facilitate trade in food and agricultural commodities. It can be used for determining microbiological mycotoxins, pesticide and veterinary drug residues, toxic metals and radio- nuclide contamination of food and feed.[16]

2.4. *Soil and Water Management and Crop Nutrition*

Nuclear techniques are used in this area to develop integrated strategies for improving the efficiency of nutrient and water use by crops within selected eco- regions, while preserving soil fertility and protecting water quality. The overall aim is to promote food security by increasing or sustaining crop productivity through better management of soil and water resources.[17]

It can be applied to increase fertility of impoverished soils thus increasing crop yields. It can also manage scarce water resources for maximising crop productivity, adding to developing of soil conservation practices for maintaining crop production and for protecting environment.[18] Nuclear techniques can also play a major role in desalinating seawater. Japan has benefited from nuclear powered desalination since the late 70s.[19]

---

[15] IAEA, 'Joint FAO/IAEA Programme, Nuclear Techniques in Food and Agriculture—Supporting Research for Agricultural Development,' Sub-Programme on 'Food and Environment Protection.'

[16] Ibid.

[17] IAEA, 'Joint FAO/IAEA Programme, Nuclear Techniques in Food and Agriculture—Supporting Research for Agricultural Development,' Sub-Programme on 'Soil and Water Management and Crop Nutrition.'

[18] IAEA, Record of the Seventh Plenary Meeting for the General Conference, 1995, p. 38.

[19] IAEA, GC (45)/INF/5, 'Nuclear Technology Review—Update 2001,' Annex 2, 29 August 2001, p. 3.

2.5. *Plant Breeding and Genetics*

Nuclear techniques are used to develop better varieties of major and under- exploited food and industrial crops. The goal is to increase food security and sustainable crop production by improving yields and quality for domestic use and export markets and by enhancing crop diversification and bio-diversity.[20]

These techniques are used to improve agronomically important characters for major crops, with emphasis on marginal and stress prone areas. They are also used to improve local neglected crop species, improvement and domestication of plant species with potential value of food as export products and speeding up the breeding of new crop varieties by increasing efficiency of mutation induction and effectiveness of mutant selection using molecular markers.[21]

2.6. *Improving Human Health*

Childhood malnutrition and micro-nutrient deficiencies remain serious global problems. In this context, nuclear techniques (isotopic tracer) have been developed for measuring whole body vitamin A under conditions of supplementation, which have now been applied in food fortification and dietary improvement programmes in several States to address problems of vitamin A malnutrition in children and pregnant or lactating women. Similarly, the assessment of iron absorption from diets to evaluate its bio-availability is an important nutritional measurement. Nuclear techniques based on stable isotopes are uniquely suited for targeting and tracking progress in food and nutrition development programmes. These techniques are used for measuring the uptake and bio-availability of many important vitamins and nutrients. Osteoporosis, cardiovascular diseases, AIDS, protein energy malnutrition, genetic disorders, diabetes and obesity are all accompanied by abnormalities in body composition. The monitoring of body composition can provide the basis for initiating preventive intervention measures. This technique has gained wide acceptance as it is inexpensive, accurate and can be applied under field conditions.

[20] IAEA, 'Joint FAO/IAEA Programme, Nuclear Techniques in Food and Agriculture—Supporting Research for Agricultural Development,' Sub-Programme on 'Plant Breeding and Genetics.'

[21] Ibid.

X-ray absorptiometry for bone density measurements is another important nuclear technique that is extensively used in the study of osteoporosis.[22]

Nuclear medicine is a clinical speciality primarily devoted to diagnostic and research applications of internally administered open sources of radioactivity. It has been widely applied because of its non-invasive nature, its ability to provide information about organ function. It also detects abnormalities at a very early stage. Today, nuclear medicine techniques are used in a broad range of medical specialities, such as oncology, endocrinology, cardiology, neurology and nephrology. There are nearly one hundred different, standardised nuclear medicine diagnostic and therapeutic techniques available for the study of the major organ systems.[23]

2.7. *Marine and Terrestrial Environment*

Nuclear techniques, being extremely sensitive and isotope specific, can dramatically complement non-nuclear methodologies for assessing pollution and particularly for understanding oceanographic processes. They can be employed to assess the environmental distributions of both radioactive and non-radioactive contaminants. Isotopic measurements using natural, anthropogenic or artificially introduced tracers can provide detailed information about complex ecosystems and the mechanisms that control the fate of diverse pollutants in the environment. In particular, isotope ratio techniques can frequently provide a unique fingerprint to facilitate source apportionment assessments. Differential decay rates can provide an insight into the time scales of oceanographic processes. It can also be used to assess human exposure to marine radioactivity.[24]

There has been much work carried out to understand the distribution, behaviour and transport of radio nuclides in the environment. The fate of radio nuclides must be understood in order to assess the possible environmental or human health consequences from unplanned releases. Extensive studies have been undertaken to investigate worldwide distribution of radionuclides from global fallout and to study the impact of re-

[22] NPT Review Conference, 'Final Document of the 2000 Review Conference of the Parties to the NPT,' NPT/CONF.2000/28, 24 May 2000, p. 56.

[23] Ibid.

[24] IAEA, 'Protection of the Marine and Terrestrial Environment.'

processing facilities, nuclear weapon test sites, radioactive waste dumping sites and nuclear accidents, including the Chernobyl accident. Global fallout from atmospheric nuclear weapon tests is, with few exceptions, the main source of anthropogenic radionuclides in the world's oceans. In some areas, however, such as the Irish and North Seas, authorised releases of radionuclides from re-processing plants have exceeded concentrations from global fallout. The Baltic and Black Seas have been the main marine reservoirs of radionuclide released after the Chernobyl accident. The Baltic Sea still has the highest marine concentration of caesium-137 in the world. (about 100 Bq/m 3)[25]

Isotope techniques have been used to determine the causes of the rising water level of the Caspian Sea. Results indicate that the responsible mechanism is increased river discharge, due primarily to changes in the hydro climatic conditions in the catchment area. Similarly, measurements of radium-226 and radon-222 have been recently used to identify a large, but previously unrecognised, submarine discharge of fresh groundwater into the oceans, resulting in better understanding and management of both marine environment and coastal aquifers. Finally, sources of terrestrial environmental contamination from nuclear testing in French Polynesia, Kazakhstan, the Marshall Islands and the Sahara Desert have also been assessed, or re-assessed, through international co-ordinated studies.[26]

2.8. *Industrial Applications*

Techniques based on radio-isotopes and radiation find widespread application in industry to improve the performance and efficiency of processes as well as the quality and the safety of industrial products. They may also confer both direct and indirect environmental benefits as compared with alternative techniques.[27]

Gamma irradiator facilities as well as electron beam accelerators are used for a number of radiation processing and irradiation applications, including curing of plastics and rubber, waste management and radiation sterilisation. Radiation processing has also produced new materials such

[25] NPT Review Conference, 'Final Document of the 2000 Review Conference of the Parties to the NPT,' NPT/CONF.2000/28, 24 May 2000, p. 65.

[26] Ibid.

[27] IAEA, 'Industrial Applications and Chemistry-Areas of Activities in Radioisotopes and Radiation Technology.'

as hydro-gel dressings for treatment of wounds. Radiation sterilisation of tissues, pharmaceuticals and pharmaceutical raw materials is growing steadily. In developed countries, almost half of the disposable medical products are sterilised using radiation. This might well increase up to 80 % with the phasing out of ozone depleting and hazardous chemical alternatives.[28]

Moreover, a large number of radio-isotopes in various chemical and physical forms are used as tracers in industrial research and development activities to check equipment performance, optimise processes, calibrate models or test installations. Gamma and neutron scanning techniques for the inspection of columns, pipes and tanks and nucleonic gauges are used for on-line measurement and process control.[29]

X-ray and gamma radiography and their complementary non-destructive testing methods are extensively used to identify defects in industrial equipment and products.[30]

The use of radiation techniques for material modification is expected to grow, particularly electron beam techniques. The petroleum, petrochemical and mineral processing industries are expected to witness an increased application of radiotracer and sealed sources technique.[31]

Non-destructive testing is likely to make greater use of digital and computer- based applications. The main focus is to improve both the efficiency and reliability of inspection. Efforts now under way will harmonise the training and certification of personnel worldwide. New developments employing micro-focus X-rays are expected to provide greater reliability and sensitivity with reduced operator dependence.[32]

2.9. *Nuclear Power Generating Electricity and Climate Change Effects*

Nuclear Power Generating Electricity

Nuclear reactors are basically machines that contain and control chain reactions, while releasing heat at a controlled rate. In electric power plants, the reactors supply the heat which turns water into steam, hence

---

[28] NPT Review Conference, 'Final Document of the 2000 Review Conference of the Parties to the NPT,' NPT/CONF.2000/28, 24 May 2000, p. 67.

[29] Ibid.

[30] Ibid.

[31] Ibid.

[32] Ibid.

driving the turbine-generators. The electricity is shipped or distributed through transmission lines to homes, schools, hospitals, factories, office buildings, rail systems and other customers.

Until now, the share of nuclear energy in generating power is not noticeably large, as the world depends mostly on fossil fuels and hydroelectric facilities. These two sources are not sustainable, while the current position of nuclear power needs to be stabilised, with possibility of future expansion. Besides, regarding nuclear power as a sustainable source of energy, public scepticism was the motive behind a projected decrease in the share of nuclear power in global electricity supply.

Worldwide, considerable efforts are being exerted to develop advanced nuclear power plants. Various organisations are involved, including governments, industries, utilities, universities, national laboratories, and research institutes. Expenditures for development of new designs, technology improvements, and the related research for major reactor types combined are estimated to exceed US $ 1.5 billion per year.[33]

Nuclear Power and Climate Change

After the environmental alert drew the attention of the international community to the impact on different aspects of the human life, part of this attention was given to energy sources. Studies were pursued to determine the effect ensued on the environment with the use of renewable or non-renewable energy sources for generating electricity, with special emphasis on climate change issues and global warming.

The debate on this matter is affecting and will continue to affect the choice of nations to opt for the use of one source of energy rather than another. It takes into consideration radioactive radiation and waste disposal, greenhouse gas emissions, land use and others.

In this regard, the effect on climate change of using nuclear energy to generate electricity, along with other elements involved in evaluating the costs of acquiring nuclear technology, will be discussed.

Among the many sources of energy, nothing has gained as much attention as nuclear energy. Despite the fact that the threat of climate change remains a serious long-term global risk, energy consumption remains one of the major sources of greenhouse gases. Nuclear power currently

---

[33] Poong-Eil, Juhn, Kuptiz Juergen and John Cleveland, 'Advanced Nuclear Power Plants: Highlights of Global Development,' IAEA Bulletin, Volume 39, Number 2, 1997.

avoids more than 8 % of the world wide carbon dioxide emissions and it meets other environmental protection objectives.[34]

On the long run, nuclear energy could replace base load fossil fuel electricity generation in most parts of the world, if generally acceptable responses can be found to meet concerns about reactor safety, radioactive waste disposal and proliferation.[35]

As for the environmental impacts, nuclear power offers specific benefits. In routine operation, nuclear power plants and fuel cycle facilities release small quantities of radioactive materials. However, the rules developed and implemented several decades ago for limiting radioactive emissions satisfy the criteria for protecting human health and are more than adequate for protecting the environment. The other emissions, residuals and burdens from nuclear power plants and fuel cycle facilities are lower than those arising from fossil-fuel electricity generation chains and are comparable or lower than those from renewable energy systems. Taking into account the entire upstream and downstream energy chains of electricity generation, nuclear power emits 40 to 100 times less carbon dioxide than the currently used fossil-fuel chains. Greenhouse gas emissions from the nuclear chain are mainly due to the use of fossil fuels in the extracting, processing, and enriching of uranium and to fuels used in the production of steel and cement for constructing reactors and fuel cycle facilities. These emissions, which are negligible, relative to those from the direct use of fossil fuels for electricity generation, can be reduced even further by energy efficiency improvements. Such improvements at the enrichment step include, for example, replacing the gaseous diffusion process by less energy-intensive processes such as centrifugation or laser isotope separation.[36]

The analysis of statistical data in different countries over the last 20 years shows that countries which implemented large nuclear programmes, such as Belgium, France and Sweden, simultaneously achieved significant reductions of their carbon dioxide, nitrogen oxide and sulphur dioxide emissions. Whereas countries and regions which have not

---

[34] Bertel, Eveline, and Joop van de Vate, 'Nuclear Energy and the Environmental Debate: the Context of Choice,' through international bodies on climate change, the roles of nuclear power and other energy options are being assessed, IAEA Bulletin, Volume 37, issue 4, 1995, p. 1.

[35] Ibid, p. 3, quoting from the executive summary of the Scientific Assessment Report prepared partially by the IAEA together with the NEA and the OECD and submitted to the Intergovernmental Panel on Climate Change.

[36] Ibid, p. 4.

employed nuclear power on a large scale, like developing countries, have a relatively high increase rate of carbon dioxide emissions.[37]

The Executive Director of the International Energy Agency in Paris, Mr. Robert Priddle, stated that though free of carbon emissions at the operational stage, nuclear power awakens grave environmental concerns. The possibility of the release of radioactive emissions, as a result of an accident or in the course of transport or storage of high-level wastes, adds to the fear that civil nuclear programmes may contribute, indirectly, to the proliferation of nuclear weapons.[38] These threats exist and will continue to exist. Although these threats have grave effects on the environment, they can be minimised with safety, good monitoring and technical solutions. However, a real dilemma may occur in case of the emission of radioactive radiation that can contaminate the environment in the far future more than any other source of energy.[39]

It can be concluded here that with the increasing demand for energy, nuclear power is one option for reducing emissions and residuals from electricity generation and for mitigating health and environmental impacts from the energy sector.

The future of nuclear power depends upon inventing new technologies that must be inherently safe, proliferation resistant and economically competitive. This technology will be capable of generating electricity at competitive prices while still satisfying regulators and investors, greater reliance on passive safety features, and passive control of nuclear materials by new fuel configurations. To meet the emerging energy needs of developing countries, these technologies must also be suited or adaptable to a broad range of environmental and industrial settings. Small and

---

[37] Ibid, p. 6.

[38] Priddle, Robert, 'Energy and Sustainable Development,' IAEA Bulletin 4, 1/1/1999, p. 5.

[39] Regarding public opinion, Mr Hans Blix, the former Director General of the International Atomic Energy Agency, in his opening statement to the International Symposium on Electricity, Health and the Environment: Comparative Assessment in Support of Decision Making stated, "However, a broad sector of public opinion in many countries—especially industrialised ones—is hesitant or opposed to an increased use of nuclear power, or even a continuation at present levels. Mainly three factors prompt these attitudes: fear of accidents, fear of long-lived radioactive wastes, and fear that use of nuclear power might contribute to proliferation of nuclear weapons. The last point is probably the least important. The expansion of nuclear power has not led to proliferation of nuclear weapons. Rather, we have seen a continuous increase in the number of countries committing themselves to non-proliferation. This factor and accelerating nuclear disarmament will probably reduce the link between nuclear weapons and nuclear power in the public's mind." Austria, 16–19 October 1995.

medium sized reactors specifically, can be a suitable choice for electricity generation—or for desalination of seawater and co-generation of heat—in remote areas or in countries with small electrical grid capacities.[40]

Another challenge affecting the future of nuclear power involves assessments of nuclear energy in relation to other energy options, in terms of factors such as environmental considerations and social effects.[41] This will be discussed in detail in the next section within the debate on the costs of acquiring nuclear technology with special emphasis on the history of using nuclear power to generate electricity.

3. *Financial and Economic Considerations for Acquiring Nuclear Technology*

Since the discovery of the peaceful uses of nuclear technology, studies concerning the costs of acquiring nuclear technology have focused on the economics of nuclear energy as a source of electricity (building reactors, acquiring raw materials and preparing them for use in the reactors, etc). These studies meant to analyse the effective costs of nuclear power in the energy market in order to compare its competitiveness with other sources of energy.

The real cost to any nation of building and operating nuclear power stations is best considered in terms of the contribution to national wealth and production which would have been obtained if the resources involved were applied to other uses. For a limited programme, in order to acquire experience or to foster export business, governments may elect to install stations that do not satisfy this criterion of economic competitiveness. They may depend, in some measure, on either taxation to cover the deficit or imposing higher charges on energy users. Very few national populations are rich enough to sustain more than a very limited burden of that kind, without suffering a serious reduction in attainable living standards. In addition, the rate of recourse to nuclear energy is no longer bound to be determined by increasing economic considerations.

In this section, the costs from the economic point of view as well as the consequence of using nuclear technology and its social implications will be highlighted.

---

[40] IAEA, 'Nuclear Power in An Era of Change,' Statement by Mr. ElBaradei, the DG of IAEA to the Pacific Basin Nuclear Conference, 2 November 2000, p. 1.

[41] Ibid, p. 2.

3.1. *The Nuclear Energy Debate*

The potential benefits of nuclear energy in the economic field have been greatly exaggerated. The maximum possible gain from the use of nuclear energy would be equal to the gain if we could get all our fuels and power gratis. That is the limit, but, in practice, since it will certainly cost a great deal to convert the necessary elements into nuclear energy, the saving will be considerably less. The cost difference between nuclear and conventional energy could not exceed the fuel cost for the latter and an upper bound on the energy cost savings that nuclear power might produce.[42] In the long run, investing in atomic electricity would be as low as the investment in ordinary thermal electricity.

In the UK, in 1953, an optimistic opinion claimed that it would be possible to produce cheap electricity by nuclear power by constructing dual-purpose reactors. However, the trend concerning the financial disadvantages of using nuclear energy started to collapse between 1953 and 1963. Opinions comparing costs with similar analyses for other public investments revealed that the benefit-cost ratios for nuclear power are hardly impressive, owing to the small ratio of economic benefits to costs. It is particularly uneconomical of public and private resources to construct nuclear plants in areas other than high-cost areas. The costs of generating nuclear power may be promising only if scientists and engineers in the nation's public and private laboratories are confident that technical advances can bring nuclear power into competitive range soon under special programs and keeping lower expenditure levels. The benefits from harnessing solar energy or developing commercially feasible method of liquefying or gasifying coal might be greater than the returns from nuclear power.[43]

The Ford Foundation Reports of 1974[44] and 1977 referred to the scientific and technological feasibility regarding both fusion and fission power. They considered that notwithstanding scientific and technological barriers, by the end of the century, controlled atomic fusion could make a large contribution to US energy needs, and it was asserted in the latter

[42] Young, Warren, 'Atomic Energy Costing,' 1998, p. 19, quoting from a document sent by Roy Harrod to Lord Cherwell for the latter's comments on 29 May 1947: Harrod to Cherwell, Cherwell Papers j 71/16, j71/15 (p. 9).

[43] Ibid, pp. 27–37.

[44] Ibid, p. 29, quoting from Ford Foundation, 'A Time to Choose: America's Energy Future,' Final Report of the Energy Policy Study Group, 1974.

report that "fusion offers the promise of practically unlimited energy. An important scientific promise has recently been made in this extremely sophisticated technology. Although it is still premature to predict success, it is believed that fusion reactors will in all probability demonstrate a useful energy output by the year 2000".[45]

The conclusion and recommendation of the 1974 Report advocated a much slower rate for nuclear power. It did not advocate an absolute ban on new nuclear plants per se. Time was needed to gain a better understanding of nuclear power problems and to reach some better judgments before major new expansions of nuclear power had been made. It stated that these decisions depended on value judgments of people which, aided by weighing benefits and risks and determined ultimately by political decisions that will depend a great deal on the people's access to information and their interest in affecting the process that generates those decisions.[46]

Support for the nuclear power increased in the 1977 Report when it maintained, "it is in the interest of a sound US energy economy to let the market establish the rate of nuclear growth. This does not mean, however, complete private ownership of the fuel cycle in few of the large capital requirements. It is not a belief that nuclear power can be treated as just another industry. Utility choices between coal and nuclear plants should be based on market considerations, within regulatory framework that deals adequately with social costs external to the industry."[47]

It summed up by stating, "in the long run of 50 years or more, the economic role of nuclear power could be more crucial. Nuclear power with the breeder and coal are the only almost certain energy sources which could provide large amounts of energy at about current costs. If fusion turns out to be unfeasible, solar energy proves too costly to exploit for electric power, and large-scale use of coal is unacceptable for environmental reasons, the economic and social importance of nuclear power and the breeder could then become substantial. We believe that some mix of coal, solar, and fusion energy, assisted by conservation, could be capable of supplying societies with long-term energy needs. But we

---

[45] Ibid, quoting from Ford Foundation 'Nuclear Power Issues and Choices,' Final Report of the Energy Policy Study Group, 1977.

[46] Yager, Joseph A., and Eleanor Steinberg, 'Energy and US foreign policy'—a Report of the Energy Policy Project of the Ford Foundation, 1974, p. 392.

[47] Young, Warren, 'Atomic Energy Costing,' 1998, p. 29.

also believe that the nuclear breeder will be competitive and is likely to eventually prove to be socially acceptable."[48]

The contradiction of the conclusions of the two reports reflects clearly the change of opinions from pessimism to optimism and from advocating a slower rate of nuclear power share to accelerating it.

All these recent reports included vague promises regarding nuclear fusion to generate electric power.

In 1982, different levels of constraints were identified concerning the development of nuclear power: the economic, the political and environmental levels. Hence, the optimistic projections made for nuclear power just before the oil crisis were based on the perceived economics of nuclear versus conventional fossil-fuel generation and expectations that electricity demand would continue to grow at historic rates. In the US, prior to 1974, both the government and the industry expected nuclear energy to be a rapidly increasing source of energy for the US economy throughout the rest of that century and well into the next. It was viewed as cheap, clean, and safe, and with the evolution of a plutonium economy, nuclear energy held the promise of providing the most economical long-run alternative to fossil fuels. At the time, the nuclear option was slowly melting away. Direct economic factors alone were not sufficient. Concerns about safety led to changes in engineering requirements for nuclear plants, increasing construction costs, delaying the licensing process and the construction process, and thereby, indirectly increasing costs. Communities presented additional concerns, as safety measures formed a stumbling barrier for utilities to find acceptable sites in many areas. Whatever the economic merits of nuclear power are, it has become a far less politically acceptable technology as a result of such safety concerns. Concerns about waste disposal and nuclear proliferation have further increased the political difficulties of getting nuclear plants approved. Nevertheless, more studies were pursued to compare the projected costs of nuclear and conventional sources of power relying on the differences in the cost of fuel.[49]

---

[48] Ibid.

[49] Ibid, pp. 54–55.

3.2. *Economics of Nuclear Energy*

Advocates claim that nuclear power offers the world a virtually unlimited source of clean and reasonably priced energy. Opponents claim that nuclear power is uneconomic, unsafe, environmentally damaging and a threat to world stability.[50]

Nuclear power stations require a substantially greater investment of capital than conventional coal or oil-fired stations. Although they resemble hydro electric plants, they differ in that the cost of the fuel to operate them is by no means negligible. In this regard, the shortage of uranium may not be a limiting factor controlling the rate of development of nuclear power. If the contracts of supplying uranium to a country expire, the lower cost of refuelling nuclear plants every four years will probably cut generating costs per unit produced. These reductions in the fuel costs however make little difference to the level of total costs because the annual financial charges covering the capital investment in nuclear plant construction are sizeable. Because of the high capital costs, a high load-factor is essential for efficient operation, and costs calculated normally assume minimum load factors of 75 or 80 per cent in order to bring the total estimated costs per kilowatt of electricity supplied as near as possible to those for conventional coal-fired stations.[51]

However, the updated NPT Technology Review of 2001 emphasised that uranium costs have dropped consistently since mid 1996. By 31 December 2000 they reached $ 18.64/kgU ($ 7.10/lb U_3O_8) compared to $ 26.52/kgU ($ 10.20/lb U_3O_8) in July 1999. For a number of years, uranium production has been about half the rate of uranium consumption, the difference being made up from uranium inventories. Substantial supplies from inventories are likely to continue in the near future and are to be supplemented by increasing significant amounts of uranium from the conversion of nuclear weapons material. New commercial production in 2000 began at the Cameco-Cogema McArthur River uranium mine in Canada in early November and at the Beverly mine in Australia about a month later. The general expectation continues to be that any upward pressure on uranium prices is likely to be modest for the immediate

---

[50] Harris, Robert, 'Alternative Energy Resources: An International Approach,' published in 'Regulation of Atomic Energy,' Columbia Journal of Transitional Law, Volume 16, 1977, p. 391.

[51] Plant, Sir Arnold, 'The Economics of Nuclear Energy' published in 'Social implications of the Peaceful Uses of Nuclear Energy,' 1972, p. 136.

future. Also, new long-term studies in 2000 judge nuclear resources to be plentiful over the longer 100-year time scales that characterise such studies. Substantial exploration and development will however be required to assure that these resources are really usable.[52]

In spite of its impressive record to date, the future of nuclear power is very much in doubt. Utilities, perhaps the strongest proponents of nuclear power, are faced with financing problems, as represented, for insurance, by the one billion dollar cost of a 1000-megawatt nuclear plant, and the problem of the time required to satisfy all licensing and environmental requirements. There is a 10–15 year lead-time for new plants.[53]

The combination of economic difficulty in financing new units, the near-future uncertainties, fuel availability and reprocessing concerns, questions related to breeder reactors and plutonium safeguard worries are frustrating and discouraging to nuclear utilities. Resolution of most of these matters falls in the government domain and must get early attention if nuclear power is to carry its share of the load as projected.[54]

From the economic standpoint, the main question is the probable cost of nuclear power as compared with power obtained from other sources. On the other hand, this cost is set against the effective demand for energy as it exists today and as is likely to arise at different locations throughout the world. Consideration of this question requires a preliminary study of the history of the relevant scientific and technological developments.[55]

3.3. *Recent Issues*

This section describes change in electricity market and the current competitive disadvantage of building new nuclear plants as well as the distinct advantages of existing ones. The costs for environmental and health effects is to be considered in view of a new generation of plants as a way to address competitiveness, safety as well as waste and proliferation concerns.

---

[52] Ibid, p. 7.

[53] Harris, Robert, 'Alternative Energy Resources: An International Approach,' 1977, p. 391.

[54] Plant, Sir Arnold, 'The Economics of Nuclear Energy' published in 'Social implications of the Peaceful Uses of Nuclear Energy,' 1972, p. 136.

[55] Ibid, p. 129.

Economic Competitiveness

During the past decade electricity supply no longer became a monopoly in the possession of the government or in the hands of a few suppliers. In some cases, it is no longer a single industry. Now, generation, transmission (over long distances), distribution (over short distances) and marketing (to final consumers) can be separate entities. Generation and marketing, in particular, are taking place in highly competitive environments where short-term pricing contracts are common, driven in part by the availability of low-cost gas generating plants for peak and base load supply.[56]

In light of these changes in the market of main electricity suppliers, the debate shifted to a new phase. The assessment of which type of power to be used to provide electricity depended upon benefits, risks and profit. The risks surrounding nuclear power are still grave: among them are environmental risks, vagueness of national rules governing licensing and the politicisation of the issue.

The competitive disadvantages influencing decisions to build new nuclear plants in countries with liberalised energy markets do not extend to existing ones, particularly when capital investments may have been depreciated or written-off during ownership transfer. Therefore, with the exception perhaps of hydro electric plants, well-managed nuclear plants, with their low fuel costs and steadily declining operation and maintenance costs, are often among the least expensive base load power to operate.[57]

On the positive side, nuclear power can offer significantly lower fuel supply and price instability risks as the required fuel quantity is small and strategic inventories for many years can be readily set up. Clearly, where indigenous fossil fuel resources are lacking, nuclear power can contribute substantially to the security of supply and stability of price as it does today in Finland, France, Sweden, the Republic of Korea and Japan.[58]

Also, in Germany, for example, an average generation cost at largely depreciated existing nuclear plants is currently approximately 30% lower than for new gas base load plants. This advantage has been sufficient to

---

[56] Ibid, p. 11.

[57] NPT Review Conference, 'Final Document of the 2000 Review Conference of the Parties to the NPT,' NPT/CONF.2000/28, 24 May 2000, p. 13.

[58] Ibid.

encourage owners of existing plants in numerous countries to invest in life extension programmes and to increase total plant generation capacity along with on-line time.[59]

Other Costs of Using Nuclear Power

Public concerns regarding the effect of using nuclear power as an alternative or a supplement to conventional sources of electricity are grave. These concerns are geared towards social effects, safety measures, waste disposal, climate change issues, land use and nuclear proliferation. Such concerns are extremely vague; nevertheless, they manage to hold the future of nuclear power at stake.

Social Costs and Health Concerns

Societies raise concerns over reactor safety, transport of radioactive materials and waste disposal. They also raise concerns over health; any dose of radiation may produce a delayed health effect. Current knowledge does not enable scientists to predict if or when these effects will occur. The delayed effects may be cancer, a thyroid disease or genetic effects. If the radiation dose is high, many cells die and immediate health effects may befall. Eventually, death may occur within days or weeks from radiation sickness. Other acute effects include vomiting and hair loss.[60]

Safety Concerns

Unquestionably, the accidents at Three Mile Island and Chernobyl have severely influenced attitudes towards safety and have damaged trust in an industry that had maintained that serious accidents were improbable or even impossible.[61] Nevertheless, the year 2000 witnessed another 400 reactor-years of safe operation added to the existing strong record of nuclear power. And, as noted above, current polls[62] and politics indicate that most people, in weighing nuclear risks and benefits, judge currently

---

59 Ibid.

60 Hanford Health Information Network, 'A Monograph Study of the Health Effects of Radiation and Information Concerning Radioactive Releases from the Hanford Site: 1944–1972,' September 1996.

61 NPT Review Conference, 'Final Document of the 2000 Review Conference of the Parties to the NPT,' NPT/CONF.2000/28, 24 May 2000, p. 9.

62 IAEA, GC (45)/INF/5, 'Nuclear Technology Review-Update 2001,' Annex 2, 29 August 2001, p. 5.

operating nuclear power plants to be acceptably safe. A radiation accident in February in Thailand served as a reminder that political acceptance of nuclear power is tied to nuclear applications beyond the nuclear fuel cycle. Polls conducted shortly after the theft of an unprotected used Cobalt-60 medical source that led to several deaths indicated the incident had undercut efforts to build support for an eventual nuclear power plant in Thailand.[63]

Improvements in safety and proliferation resistance are central objectives for some 25 innovative reactor designs in various stages of design and development around the world. To reinforce different national initiatives, the IAEA launched a new International Project on Innovative Nuclear Reactors and Fuel Cycles. In its first phase, work will proceed in five subject areas recognised as important for future development of nuclear energy technology. The five subject areas are: Resources, Demand and Economics, Safety, Spent Fuel and Waste, Non-proliferation and Environment. This will involve the selection of criteria and development of methodologies and guidelines for the comparison of different concepts and approaches, taking into account the compilation and review of such concepts and approaches and the determination of user requirements in the subject areas. It was concluded that a better understanding of the human role in plant operation was needed. The industry responded through technological safety improvements, such as updated control room layouts that provide operators with clear and necessary information and improved training and operational procedures, along with exacting internal and external audits of operating and safety performance.[64]

Waste Concerns

The lack of adequate information is a factor that allows the radioactive waste issue to remain a public concern. A recent opinion survey carried out by the EU of approximately 1000 people in each Member State showed that only 20 % considered themselves well informed about radioactive waste, 44 % considered themselves poorly informed and 32 % very poorly informed. Roughly 50 % thought nuclear waste was disposed of at sea or sent to other countries.[65]

---

[63] NPT Review Conference, 'Final Document of the 2000 Review Conference of the Parties to the NPT,' NPT/CONF.2000/28, 24 May 2000, p. 6.

[64] Ibid.

[65] Ibid, p. 9.

Although the scientific community is generally convinced that disposal is technically achievable, the reality that waste is accumulating and disposal has not yet taken place leads to the perception that it cannot be done.

Land Use

The production and subsequent supply of electricity has and will continue to require the use of land and, therefore, has an impact on the amount of land available for other purposes. It can be argued that such impacts are likely to be reversible in the future but practical experience, such as the US superfund activities to reclaim contaminated land, has shown that even with existing technology, the resources and political will to address such problems may be lacking. The impact on land is probably one of the most contentious and socially significant impacts need to be considered in energy policy decision-making. There may be significant social costs in the far future.[66]

It is known that land contamination caused by conventional fuels is likely to have less impact than uranium. Effects of oil spills, for example, tend to disappear after several decades. But that is not the case with nuclear radioactive materials; their effects can last for thousands of years.[67]

3.4. *Environmental Effects (Climate Change Debate)*

Environmental Effects

As for environmental concerns, the use of nuclear power has proved to be environmentally sound with regards to the emission of greenhouse gases. But the apprehension lies in the grave environmental effects that may occur as a result of an accident. Some scientists argue that the same effect of reducing carbon dioxide emissions would be feasible without nuclear power if renewable energy sources would enter into the market to be deployed at very high rates during the next century. However, there are hazards other than gaseous emissions that should be taken into account when conducting comparisons. Controversy surrounds any large-scale hydro-electric project, in terms of population displacement,

---

[66] Rabl, Ari, and Mona Dreicer, 'An Eye to the Future—Long-Term Health and Environmental Impacts on Energy Systems,' IAEA Bulletin 4, 1/1/1999, p. 33.

[67] Ibid.

loss of habitat, and change in the watercourse and so on. Some of the materials used in the production of photovoltaic cells are toxic and hazardous. Geothermal operations can release heavy metals that may leach into groundwater. To produce energy from energy crops requires well over 100 times the land area required to produce the same electricity from coal. Wind farms have an undesirable visual impact, can generate irritating noise and interfere with electromagnetic communications.[68]

Climate Change Debate

After the year 2000, the Kyoto Protocol, which aims at reducing greenhouse gas (GHG) emission, did not enter into force until 2005. In implementing the protocol there remains a distinct possibility that, for two of its three flexibility mechanisms (the Clean Development Mechanism and Joint Implementation), its members will penalise GHG reductions from nuclear power relative to reductions from renewables or efficiency improvements. But two considerations suggest that nuclear power plants may yet be able to turn their near-zero GHG emissions into a clear economic credit on the bottom line. First, many businesses are ahead of government negotiators in terms of hedging against, or perhaps betting on, a future carbon-constrained business environment. Such firms (including, for example, BP and Shell) are generating internal GHG reduction markets, an increasing volume of completed trades, and a growing number of ancillary services in brokering, clearing and auditing trades. As these markets grow, so should the opportunities for nuclear power plants to benefit economically from their low GHG emissions. Second, several countries are, or will be implementing GHG reduction policies independent of near-term progress on the Kyoto Protocol. Some implementation policies (such as "green certificate" programs) might be applied in ways that penalise nuclear power relative to renewables, but others (e.g., carbon taxes or tradable carbon emission permits) should create a direct economic advantage for nuclear power.[69]

---

[68] Priddle, Robert, 'Energy and Sustainable Development,' 1999, pp. 4–5.

[69] NPT Review Conference, 'Final Document of the 2000 Review Conference of the Parties to the NPT,' NPT/CONF.2000/28, 24 May 2000, p. 8.

4. *Future of Nuclear Energy*

No technology is risk-free and it is essential for the future of nuclear power that technological innovations continue to increase levels of safety and to reinforce and extend the safety culture that has developed within the industry. Firm opponents of nuclear power argue that the risks of the technology outweigh its benefits, relative to alternatives. However, polls in 2000[70] and general political trends indicate that the majority of people judge benefits to outweigh risks, at least for the current operation of nuclear power plants. Looking to the future, there is a debate whether nuclear power can be classified as sustainable technology. The World Energy Assessment (WEA), published in September 2000, suggests in part that while technologies and management strategies necessary for nuclear sustainability may be possible, the decisions will be largely political, rather than technical and economic. The WEA report (like others) suggests that to improve its political prospects, the nuclear power industry should better address remaining concerns about waste, safety and proliferation.[71]

In holding the early cost comparisons between nuclear and coal-based power and their impact on the nuclear power program, there was a fundamental tug of war between 'as safe as possible' and 'as cheap as possible'. The economics of early reactors was decidedly unfavourable. If nuclear power was to compete, the designers knew that they would have to look for ways to cut costs. Before Chernobyl, most citizens opted for the cheapest possible electricity. Government regulators were eager to satisfy this popular demand. Electric utility executives were less interested in generating power in what seemed to them to be the most economical way. It is hardly surprising, therefore, that the economic issue has played a central role in the nuclear controversy. But the information doesn't exist to resolve the question one way or another and will not exist for several years to come. In these circumstances, it seems best to proceed on the assumption that all of us are interested in cheap electricity, in a narrow micro-economic sense, either nuclear or coal-fired power.[72]

---

70 IAEA, GC (45)/INF/5, 'Nuclear Technology Review-Update 2001,' Annex 2, 29 August 2001.

71 NPT Review Conference, 'Final Document of the 2000 Review Conference of the Parties to the NPT,' NPT/CONF.2000/28, 24 May 2000, p. 5.

72 Young, Warren, 'Atomic Energy Costing,' 1998, pp. 94–.95.

The upshot is that whenever there is easy access to coal and oil, new conventional-type stations will continue, at least in the near future, to be more advantageous than nuclear power. Nuclear stations will make their greatest impact in regions remote from coal and oil but with a heavy continuous demand for power, justifying a large-scale installation. Some may be regions that have become highly industrialised on account of a variety of location advantages that outweigh the relatively high cost of power. They may, for example, serve a large but localised market for particular products which on a balance of considerations is best catered to by local factories; and the development of these areas may favour new diversification for which cheaper power is essential. Other regions may be plentifully supplied with valuable materials, such as minerals. They lose a large proportion of their crude weight or bulk at early stages in the manufacturing process. They are therefore better processed near the source of supply than transported in crude condition to other regions where power is much cheaper. The rate of industrial development of these areas is limited by the high cost of power, due to high transmission losses on electricity generated in other areas where conventional fuels are more abundant. If their prospective demand for power has the appropriate characteristics—an adequate volume to justify a large installation and a sufficiently steady and high load-factor to permit continuous operation under optimum conditions—a local nuclear plant may successfully provide cheaper electricity for further growth better than any other source. In the foreseeable future, a 12% financial charge would be economical only if the development potential were exceptionally great and other fuels abnormally dear. Where the interest rates are high, the sacrifice to be sustained by the population, in terms of the contemporaneous deprivation of alternative economic and social amelioration, would be a heavy burden for which only the most exceptional promise of ultimate benefit from nuclear power would afford an adequate compensation.[73]

The need for public acceptance of nuclear technology as a source of energy and a means to addressing human problems in a positive manner is crucial.

Until the concerns about safety, environmental effects, waste disposal, land contamination, health risks and nuclear proliferation are brought to an acceptable political resolution, it is safe to assume that the nuclear-energy industry will grow at a snail's pace, if at all. Economic competi-

[73] Plant, Sir Arnold, 'The Economics of Nuclear Energy' published in 'Social implications of the Peaceful Uses of Nuclear Energy,' 1972, p. 138.

tiveness is not enough to address the social problems that may occur as a result of using nuclear power. The vagueness surrounding the peaceful uses of nuclear energy and the huge publicity of the horrible effects of using a nuclear weapon formulated a hostile public opinion internationally. This opinion will not be demolished over night; it requires work, time and hard efforts from scientists, economists and others to enlighten the public with the benefits of using nuclear technology in their lives and to address efficiently the problems concerning use of this technology.

CHAPTER TWO

TRANSFER OF NUCLEAR TECHNOLOGY FOR PEACEFUL PURPOSES UNDER INTERNATIONAL LAW

1. *Introduction*

Since the dawn of nuclear discovery, and the revelation of the benefits of the atom, countries have attempted to acquire nuclear capabilities. However, the military misuse or abuse of the atom has always posed a threat to the existence and safety of humanity. To balance the situation, the international community sought to establish a system to encourage the use of nuclear technology for peaceful purposes while simultaneously ensuring the non-proliferation of nuclear weapons.

During the Second World War, the US decision to impose tight secrecy on atomic energy dampened, for at least a decade, any possibility of cooperation in the development of nuclear energy for peaceful purposes. Later, the Eisenhower Atoms for Peace Programme in 1953 and the declassification of large amounts of nuclear information in 1955 at the first UN Geneva Conference on Peaceful Uses of Atomic Energy paved the way to exploit this knowledge. Encouraged by growing public enthusiasm, Canada and Sweden joined the US, the UK, the former USSR and France in developing nuclear reactor models. By 1964, there were fifteen reactors in operation or completed, using either natural or enriched uranium fuel and the sums thus invested were, already, considerable.[1]

2. *Atoms for Peace*

At the UN, on 5 December 1953, President Eisenhower officially launched the "Atoms for Peace" plan, which included the following principles:[2]

---

[1] Nuclear Energy Agency of the Organisation for Economic Co-operation and Development, 'The Regulation of Nuclear Trade: Non-Proliferation—Supply—Safety: National Regulations,' Volume 1, 1988, p. 11.

[2] Ibid, p. 13.

- Separating civilian and military uses of atomic energy through effective safeguards and inspections.
- Imposing an embargo on the entire fuel cycle, except for certain sensitive nuclear material and equipment.
- Imposing control over any sensitive material and equipment that is transferred to NNWS.

In 1955, the UN-sponsored Geneva Conference on the Peaceful Uses of Atomic Energy laid the basis for establishing the IAEA in 1957 to promote the transfer of nuclear technology for peaceful purposes and to supervise the transferred technology in a manner to prevent its diversion to military purposes, as emphasised in Article II of the IAEA Statute.

3. *Adoption of NPT*

The establishment of the IAEA and the development of specific safeguards were not sufficient to prevent the non-proliferation of nuclear weapons. Consequently, the international community crowned its efforts to ensure the non-proliferation by concluding the NPT in 1968, which entered into force in 1970 and was extended indefinitely in 1995 under certain conditions that will be elaborated in this chapter. The NPT became the main pillar in the international nuclear non-proliferation regime. It did not establish an organisation to ensure the compliance of the NPT member States with their obligations under the Treaty, but instead entrusted the IAEA with this job.

In addition, the IAEA and the NPT recognised the role of regional and bilateral agreements in promoting the transfer of nuclear technology for peaceful purposes and encouraged it.

This chapter will discuss the following:

- Transfer of Nuclear Technology under NPT rules.
- Transfer of Nuclear Technology under IAEA rules.
- International Cooperation.

*

SECTION ONE
TRANSFER OF NUCLEAR TECHNOLOGY UNDER NPT RULES

1. *Introduction*

Due to the gravity of the spread of nuclear explosives, NWS (US, Russia, UK, France and China)[3] sought to preserve their own nuclear arsenals, and be free to develop them, but at the same time, to prevent other countries from acquiring nuclear weapons via a binding legal instrument.

To this end, NWS were obliged to make some promises in return. They promised to transfer nuclear technology for peaceful purposes to NNWS; provided that the latter's nuclear facilities would be kept under international supervision. They also promised to take sincere steps towards reducing their nuclear arsenals until they were fully eliminated. The NPT is the outcome of a bargain and a compromise. Although the main goal of the NPT is to create a world free of nuclear weapons where nuclear technology is used only for peaceful purposes for the benefit of humankind, it is, in fact, a discriminatory treaty. It divides the world into two classes of States, those that have nuclear weapons and those that do not.[4]

The NPT established a dual-faced system; it supports the development of peaceful uses of nuclear technology and simultaneously imposes a disarmament obligation on both NWS and NNWS. NNWS are obligated not to receive a nuclear explosive device or any assistance in manufacturing one, and to keep their nuclear facilities dedicated to peaceful purposes under IAEA safeguards. NWS are obligated not to transfer nuclear material or equipment to an NNWS that may assist in any way in the manufacture of a nuclear explosive device. They are also obligated to engage in negotiations in good faith to reduce their nuclear arsenals until the ultimate goal of the treaty is fulfilled and nuclear weapons are completely eliminated (Article VI of the NPT). The 1995 Review and Extension Conference

[3] The ICJ in its Advisory Opinion on the 'Legality of the Threat and Use of Nuclear Weapons' emphasised in paragraph 61 that only the five nuclear powers have the right to keep these weapons according to the NPT, it stated, inter alia, "… the possession of nuclear weapons by the five weapon States has been accepted." And according to the NPT, the NWS are China, France, Russia, UK and US. These are the States which had conducted a nuclear explosion before 1967 and they are the only states which possess nuclear weapons legally according to the Treaty.

[4] UN, 'Assuring the Success of the Non-Proliferation Treaty Extension Conference,' Excerpts from the panel discussions organised by the NGO Committee on Disarmament, Inc., at the Conference held at the United Nations in New York, 20–21 April 1994, p. 107.

reiterated that it is important to reach "the ultimate goals of the complete elimination of nuclear weapons" and to conclude "a treaty on general and complete disarmament under strict and effective international control ..."[5]

The 1995[6] and 2000[7] Review Conferences of the NPT reaffirmed the main objectives of the Treaty, when they recognised the following principles:

- The transfer of nuclear technology should be consistent with States' obligations under the treaty.
- States party to the treaty should not cooperate with or assist States not party to it in the nuclear field or export nuclear related dual-items that may assist in any nuclear weapons programme.
- States party should not provide source or special fissionable material or equipment to any NNWS unless they are subject to safeguards.[8]

The NPT codified a norm against the legitimacy of acquiring nuclear weapons; a legal basis for export controls and for international safeguarding of nuclear materials; and also a legal basis for imposing sanctions in cases of violation.[9]

2. *Obligation to Transfer Nuclear Technology for Peaceful Uses*

Article IV of the Treaty addresses the inalienable right of all Parties to develop research, production and use of nuclear energy for peaceful purposes without discrimination and declares that they will undertake to facilitate, and will have the right to participate in, the fullest possible exchange of equipment, materials and scientific and technological information for peaceful uses of nuclear energy. It has also encouraged con-

---

[5] NPT Review and Extension Conference, 'Decisions and Resolutions Adopted by the 1995 Review and Extension Conference of the Parties to the NPT,' New York, 17 April–12 May 1995, Appendix A-2, p. 278.

[6] Ibid, pp. 278–279.

[7] NPT Review Conference, NPT/CONF.2000/28, 'Final Document of the 2000 Review Conference of the Parties to the NPT,' 24 May 2000, p. 2.

[8] NPT Review and Extension Conference, 'Decisions and Resolutions Adopted by the 1995 Review and Extension Conference of the Parties to the NPT,' New York, 17 April–12 May 1995, p. 279.

[9] Schienman, Lawrence, 'The Role of Multilateral Regimes in the Non-Proliferation,' Transnational Law and Contemporary Problems, Volume 2, Number 1, 1992, pp. 569–586, 573 f.

tributing States party to the treaty to co-operate in contributing alone or together with other States or with international organisations to transfer nuclear technology for peaceful purposes, especially in the territories of NNWS party to the Treaty, with due consideration for the needs of the developing areas of the world.

In analysing Article IV, the following will be addressed:

1. Encouraging peaceful uses of nuclear energy.
2. The inalienable right to develop research, production, and use of nuclear energy for peaceful purposes.
3. Balance of Obligations and Rights in Article IV.

2.1. *Encouraging Peaceful Uses of Nuclear Energy*

The preamble of the NPT affirmed, inter alia, "the principle that the benefits of peaceful applications of nuclear technology, including any technological by-products which may be derived by NWS from the development of nuclear explosive devices, should be available for peaceful purposes to all the parties to the Treaty, whether NWS or NNWS." It stated further that, "in furtherance of this principle, all Parties to the Treaty are entitled to participate in the fullest possible exchange of scientific information for, and to contribute alone or in cooperation with other States to, the further development of the applications of the atomic energy for peaceful purposes ..."

- The peaceful uses of nuclear energy received attention at the consecutive NPT Review Conferences starting with the 1975 Review Conference,[10] and ending with the 1995 Review and Extension[11] and the 2000 Review Conference.[12] The following principles were therewith highlighted:
- Attaching importance to ensuring the exercise of the inalienable right of all Parties to the Treaty to develop research, production and use of nuclear energy for peaceful purposes without discrimination, as it constitutes one of the fundamental objectives of the Treaty,

[10] NPT Review Conference, 'Final Declaration of the [first] Review Conference of the States Parties to the NPT,' May 1975.

[11] NPT Review and Extension Conference, 'Decisions and Resolutions Adopted by the 1995 Review and Extension Conference of the Parties to the NPT,' New York, 17 April–12 May 1995, p. 279.

[12] NPT Review Conference, NPT/CONF.2000/28, 'Final Document of the 2000 Review Conference of the Parties to the NPT,' 24 May 2000, p. 10.

- States' decisions in the field of peaceful uses of nuclear energy should be respected without jeopardising their policies or international cooperation agreements and arrangements for peaceful uses of nuclear energy and its fuel cycle policies,
- Fostering the development of peaceful uses of nuclear energy by providing a framework of confidence and cooperation within which those uses can take place,
- Parties to the treaty should facilitate, and have the right to participate in, the fullest possible exchange of nuclear technology for peaceful uses among States parties to the treaty. This contribution can help States to overcome the technological and economic disparities between developed and developing countries,
- Preferential treatment be given to NNWS Party to the treaty, taking into account the needs of developing countries,
- Enhancing peaceful uses of nuclear energy by all parties and cooperation among them,
- Parties to the treaty need to discuss regularly and take specific steps towards the implementation of Article IV of the Treaty.

2.2. *The Inalienable Right to Develop Research, Production and Use of Nuclear Energy for Peaceful Purposes*

During different phases of the NPT negotiations, there was unanimity that the Treaty should certainly not hamper full access to knowledge and technology in the field of peaceful uses of nuclear energy. This right is equally significant to developed as well as developing countries. The exercise of this right is subject to two conditions: non-discrimination and conformity with Articles I and II of the NPT to develop research, production and use of nuclear energy for peaceful purposes. This requires the observance of the restrictions imposed in both articles on the transfer and acquisition of nuclear devices.[13]

The NPT Review Conferences, as previously stated, reiterated that nothing in the NPT had been identified as affecting the inalienable right prescribed in Article IV.1.

In practice, the nuclear supplier States have imposed more restrictions affecting the full exercise of the inalienable rights. (This will be elaborated

[13] Shaker, Mohamed I., 'The Nuclear Non-Proliferation Treaty: Origin and Implementation 1959–1979, the Treaty on the Non-Proliferation of Nuclear Weapons,', 1980, pp. 334–339.

upon in Chapter 3). Also, the case of Iran (which will be taken up in the second case study of this book) exposed the inadequacy of the NPT rules to provide for means to enforce this right. Although the wording of Article IV stipulates in a clear and direct manner that NNWS have an inalienable right to access nuclear technology, implementation revealed that the treaty does not provide for a system that brings this right into effect in practice.

2.3. *Balance of Obligations and Rights in Article* IV

Article IV has created a balance between the following:

i. Fullest possible exchange of equipment, materials and scientific and technological information.
ii. Cooperation to develop applications of nuclear energy for peaceful purposes.

Fullest Possible Exchange of Equipment, Materials and Scientific and Technological Information

The 1995 Review and Extension Conference called for encouraging cooperation in the exchange of equipment, materials and scientific and technological information. It has called upon all State parties to observe their legitimate right, particularly developing States, to full access to nuclear material, equipment and technological information for peaceful purposes.[14]

Obligation to Facilitate and the Right to Participate in the Fullest Possible Exchange

Article IV created a balance between the rights of NNWS and the commitment to action by NWS as well as by others in a position to contribute.

The 1995 Review and Extension Conference highlighted several rules relating to this obligation, among which are the following:[15]

- States should undertake to facilitate participation in the fullest possible exchange of equipment, materials and scientific and technological information for the peaceful uses of nuclear energy.

---

[14] NPT Review and Extension Conference, 'Decisions and Resolutions Adopted by the 1995 Review and Extension Conference of the Parties to the NPT,' New York, 17 April–12 May 1995, p. 279.

[15] Ibid.

- States should give preferential treatment to NNWS party to the treaty in all activities designed to promote peaceful uses of nuclear energy.
- The needs of developing countries should particularly be taken into account.
- Promoting transparency in nuclear-related export controls.
- Encouraging the IAEA efforts aimed at finding ways and means for funding technical assistance through predictable and assured resources.

Exchange of Equipment and Materials

Since the adoption of the NPT, NNWS have raised concerns related to their access to raw materials, modern science and technology. These concerns remained as an issue of controversy from the Conference of NNWS in 1968 until the last NPT Review Conference in the year 2000. NNWS also raised concerns about restrictive measures enforced by nuclear suppliers that go beyond the requirements of the NPT. (This will be discussed later in Chapter 3).

Exchange of Scientific and Technological Knowledge

The NPT encouraged its parties to cooperate in this area and to exchange scientific and technological knowledge. However, in practice, concerns were raised regarding this exchange, especially regarding sensitive techniques relevant to enriching uranium and building power reactors.

The IAEA plays an important role in disseminating scientific knowledge as described in Chapter 1. Nonetheless, the IAEA itself is restricted by its limited resources such that developments in the nuclear field take place in NWS and the developed NNWS more than in developing NNWS.

The trend that is followed by NWS and developed NNWS is to keep their monopoly over the new developments in the nuclear field.[16] They do not provide other States with such developed knowledge thereby contradicting the spirit of Article IV of the NPT that gives NNWS party to the treaty the inalienable right to access scientific and technological knowledge.

[16] Statement of Raouf, Tarik, Section Head, Verification and Security Policy Coordination, Office of External Relations and Policy Coordination at the IAEA, in an interview concluded at IAEA Headquarters, 2004.

Cooperation to Develop Applications of Nuclear Technology for Peaceful Purposes

Article IV of the Treaty included the right of parties "to contribute alone or in cooperation with other States to the further development of the applications of nuclear energy for peaceful purposes with due consideration for the needs of the developing areas of the world."

Contributing States

Article IV introduces a new category of States, broader than NWS and narrower than NNWS as "Contributing States". The obligation derived from this Article falls upon the "Contributing States," i.e., suppliers, be they NWS or NNWS, towards beneficiaries, i.e., recipient states (especially developing countries).[17]

It is difficult to identify "Contributing States". They can be States that are members of the Supplier Groups. However, such is not a very accurate criterion because there are other developed States in the nuclear field that are not members of these groups like India, Pakistan and Israel. However, its members are adequately developed in the nuclear industry and can be considered contributors.

Beneficiary States

Article IV stipulates that beneficiaries are NNWS Party to the Treaty and it provided special consideration for the needs of the developing areas of the world.

All States can be "Beneficiary States" even highly developed NWS can be recipients to nuclear material. The less developed NNWS can also be beneficiaries or recipients to nuclear material and equipment, to scientific and technological knowledge and services affiliated to running nuclear facilities. The "Beneficiary State" can either be a party to the NPT or not. However, the NPT requested "Contributing States" to give weight to the adherence to the Treaty by "Beneficiary States".[18]

Membership in the NPT did not give much leverage. Some of its members were not able to set off enough channels for cooperation for peaceful purposes while others were able to have full cooperation and assistance, like Israel, Pakistan and India. They are not members of the

---

[17] Shaker, Mohamed, 'The Nuclear Non-Proliferation Treaty: Origin and Implementation 1959–1979,' 1980, p. 317.

[18] Ibid, p. 320.

NPT, but they had sufficient cooperation with nuclear suppliers to the extent that it enabled them to build nuclear weapons.

Obligation of Contributing States

States that are members of the NPT are requested to provide assistance in all peaceful nuclear fields to NNWS members to the Treaty. This assistance includes exchange of knowledge, building nuclear research reactors and nuclear power plants. This assistance includes commercial transactions as well.

The main preoccupations of advanced NNWS include the supply of nuclear fuel information on uranium enrichment techniques and reactor construction. Generally speaking, developing countries are more in need of economic and technical assistance, training, and capital investments. In 1969, these needs were pointed out by GA as follows:[19]

- On the transfer of nuclear technology, there is a need for more nuclear centres to form a valuable link in such transfer. When a strong cadre of research scientists is available, consideration may be given to nuclear research reactors.
- On nuclear minerals, more intensive exploration for uranium is an important means in which international assistance could be provided to developing countries.
- Encouraging manufacturers to take a greater interest in medium-sized nuclear power plants that are more suitable for the needs of a growing number of developing countries.
- For the remote future, agro-industrial complexes are considered promising.
- On radio-isotopes and ionising radiation, their effects could benefit developing countries.
- Technical cooperation projects carried out by IAEA would remain the chief source of assistance in introducing nuclear science and technology into many of the developing countries. However, the tight financial position of the IAEA remains a concern and there should be a steady increase in the resources available for multilateral technical assistance.

The 2000 Review Conference of the NPT acknowledged the importance of the work of the IAEA, as the principal agent for technology transfer among international organisations, and it affirmed the importance of technical

[19] Ibid, p. 335.

cooperation activities of IAEA, as well as bilateral and other multilateral cooperation, in fulfilling the obligations set forth in Article IV of the treaty.[20]

3. *Obligation to Share Peaceful Applications of Nuclear Explosions*

Article V mainly addresses NWS to share the potential benefits from any peaceful applications of nuclear explosions with NNWS party to the treaty on a non-discriminatory basis. It also comprises the aspect of the peaceful uses of nuclear energy to which Article IV equally applies.[21]

This Article forms a commitment to ensure that NNWS share the benefits derived from peaceful uses of nuclear explosions. However, as a consequence of the cessation of nuclear tests, after the conclusion of a Comprehensive Nuclear Test Ban Treaty (CTBT) in 1996, this Article will have no future application other than sharing old information that might have been kept from NNWS.

4. *Non-Diversion and Disarmament Obligation under the* NPT

Disarmament obligations under the NPT are generally considered of particular consequence. By signing the NPT, Parties agreed that self-imposed arms denial of one side—NNWS—was to be matched, ultimately, by corresponding acts of the other side—NWS. The parties have therefore undertaken to pursue negotiations 'in good faith' to halt the nuclear armament. As a fact, the NPT is the only existing international legal document under which NWS are specifically committed to nuclear disarmament.[22] Moreover, this commitment is loose. NWS maintain that this obligation remains as a promise. The treaty did not specify any time frame to halt nuclear armament. There is no monitoring system similar or equivalent to safeguards on NNWS, and there are no assurances that NWS will ever live up to their obligations under Article VI of the NPT.

[20] NPT Review Conference, NPT/CONF.2000/28, 'Final Document of the 2000 Review Conference of the Parties to the NPT,' 24 May 2000 p. 14.

[21] Shaker, Mohamed, 'The Nuclear Non-Proliferation Treaty: Origin and Implementation 1959–1979,' 1980, p. 382.

[22] Goldblat, Josef, 'Twenty Years of the NPT implementation and prospects,' 1990, p. 37.

4.1. *Non-Diversion Obligation on NNWS*

According to the NPT, NNWS enjoy an inalienable right to benefit from exploiting peaceful uses of nuclear technology. On the other hand, they are obligated not to divert nuclear energy from peaceful uses to nuclear weapons or other nuclear explosive devices (Article IV). In addition, NNWS are required to refrain from receiving, gaining control of, manufacturing, acquiring, seeking or receiving assistance to manufacture nuclear weapons or other nuclear explosive devices. (Article II).

This is a negative obligation of disarmament on NNWS. They are not required by the treaty to take any active action. They are only obligated to desist from pursuing nuclear weapons.

4.2. *Disarmament Obligations on NWS*

Article VI confirms the main purpose of the treaty, as stipulated in the preamble, that the parties to the treaty declare 'their intention to achieve at the earliest possible date the cessation of the nuclear arms race and to undertake effective measures in the direction of nuclear disarmament.'[23]

It raises several obligations on all parties to the treaty, but NNWS watch NWS as the main addressees by this Article. Their obligations include:

i. Cessation of the nuclear arms race.
ii. Nuclear disarmament.

Cessation of the Nuclear Arms Race

Article VI obligates NWS to achieve the 'cessation of nuclear arms at an early date'. To this end, three measures were stipulated in disarmament forums: a comprehensive ban on nuclear tests, thereby banning production of fissionable materials for nuclear weapons and prohibiting the use of nuclear weapons. So far, none of these measures has been achieved in full; only the efforts to ban nuclear tests have finally become successful with the conclusion of a CTBT in 1996. Article I of the treaty provides that the main objectives of the treaty is to obligate states parties not to carry out any nuclear weapon test explosion or any other nuclear explosion, to prohibit and prevent any such nuclear explosion at any place under

[23] Bunn, G, R.M. Timberbaev, and J.F. Leonard, 'Nuclear Disarmament: How much have the five nuclear powers promised in the NPT?', June 1994, p. 2.

its jurisdiction or control, and to refrain from causing, encouraging, or in any way participating in the carrying out of any nuclear weapon test explosion or any other nuclear explosion.

As for the prohibition of the production of fissionable materials, in 1994, the Conference on Disarmament (CD) appointed a special coordinator to consult other parties on the scope and forum for negotiating 'a non-discriminatory, multilateral and internationally and effectively verifiable treaty banning the production of fissionable material for nuclear weapons and other nuclear explosive devices.'[24]

Meanwhile, the situation remains stagnant regarding the prohibition of use of nuclear weapons. Since the adoption of NPT, the only measure that NWS have undertaken was issuing unilateral declarations promising not to use nuclear weapons against NNWS and not to be the first to use them against other NWS.[25]

It can be concluded that until this date, little has been done to ensure the cessation of the nuclear arms race around the globe.

Nuclear Disarmament

Since signing the NPT, only the US and the former USSR have concluded bilateral agreements on the limitation and reduction of strategic ballistic missiles and defences against their use as a step towards implementing Article VI. But, the UK, France and China are not engaged in negotiations to limit the development of nuclear weapons or their means of delivery.[26]

The aforementioned steps undertaken by the US and the former USSR are in compliance with Article VI. Although the pace seems too slow to achieve the main goal of the NPT, they could eventually lead to nuclear disarmament. Criticism is still directed towards this slow pace or against the other three NWS who have yet to engage in any negotiations to reduce their nuclear arsenals and the question still remains if they will ever commence these negotiations!

Notwithstanding the fact that the measures undertaken by NWS to comply with Article VI are minimal, they argue that they have complied with the terms of the NPT and the treaty has proved its value. On the contrary, NNWS argue that the measures undertaken by NWS to achieve

24 Ibid, p. 2.

25 Goldblat, Josef, 'Twenty Years of the NPT Implementation and Prospects,' 1990, pp. 37–38.

26 Ibid, p. 38.

the utmost goal of the NPT, namely, to reach general and complete disarmament are not sufficient. These positions have been widely rehearsed during the review conferences and their preparatory committees. In the Preparatory Committee of the 2005 Review Conference of the NPT, held at the UN in NY in 2004,[27] the US reiterated that it has reduced its nuclear arsenal; hence, it has complied with its obligations under Article VI.[28] However, the Nuclear Posture Review (NPR) prepared by the Department of Defence in December 2001, stressed the importance of the role played by nuclear weapons as a strategic deterrent in the US defence policy and it has emphasised the importance of developing new tactical nuclear weapons. This contradicts the measures taken towards reduction. The US demolishes part of its old stockpiles, while subsequently substitutes them with new ones, thereby constituting a violation of its obligations under Article VI of the NPT to negotiate to reduce and eventually eliminate its nuclear weapons, not to develop new models of the weapon.[29]

On the other hand, the 1995 Review Conference of the NPT adopted a decision regarding "Principles and Objectives for Nuclear Non-Proliferation and Disarmament," and the 2000 Review Conference agreed to adopt practical steps for systematic and progressive efforts to implement Article VI so as to reach the ultimate objective of the disarmament process is general and the complete disarmament under effective international control.[30]

The 1995 Review and Extension Conference discussed the implementation of Article VI where NWS reaffirmed their commitment, as stated in Article VI, to pursue negotiations in good faith on effective measures relating to nuclear disarmament. It has also laid out a programme of action to ensure effective implementation of Article VI, including the immediate commencement and early conclusion of a convention banning the production of fissile material for nuclear weapons or other nuclear explosive devices and the determined pursuit of systematic and

[27] NPT, 'Preparatory Committee for the 2005 Review Conference of the Parties to the NPT,' Third Session, New York, 26 April–7 May 2004.

[28] UN, 'Assuring the Success of the Non-Proliferation Treaty Extension Conference,' 1994, p. 12; this point of view was expressed by Thomas Graham from the US Army Control and Disarmament Agency.

[29] US Department of Defense, 'Nuclear Posture Review 2001,' Annual Defense Department Report, 31 December 2001, p. 16.

[30] NPT Review Conference, NPT/CONF.2000/28, 'Final Document of the 2000 Review Conference of the Parties to the NPT,' 24 May 2000, p. 18.

progressive efforts by NWS to reduce nuclear weapons globally, and by all States to reach general and complete disarmament under strict and effective international control.[31]

The NPT as it stands has not managed to prevent proliferation, but the conclusion of a CTBT is significant in assisting the NPT to achieve its goals. However, a strong CTBT will not end proliferation either. NWS, parties or not to either the NPT or the CTBT, have not and will not give up their capability to modernise and design nuclear weapons through computers and simulations. The cases of India and Pakistan mirror this reality. They were able to acquire nuclear weapons, they conducted nuclear tests and they are still developing their nuclear capabilities without any restrictions. Israel too is still outside the NPT and has nuclear weapons and is developing them at will.

Most, if not all, NNWS joined the NPT hoping for the implementation of Article VI. This was the base of the compromise they had accepted to remain non-nuclear. Nevertheless, the lack of a time table to eliminate nuclear weapons and the continuity of the current situation will lead to weakening the treaty.[32]

5. *Impacts of the NPT Indefinite Extension*

In 1995, according to Article XI, the NPT was extended indefinitely, 25 years after its entry into force. The indefinite extension did not assist in removing threats to the non-proliferation system; but it has rather ensured the survival of the current one with the following flaws:

1. It gives NWS indefinite time to fulfil their obligation under Article VI of the treaty, "to pursue negotiations in good faith on effective measures relating to cessation of the nuclear arms race at an early date and to nuclear disarmament, and on a treaty on general and complete disarmament under strict and effective international control." There is a persistent need for a time frame to eliminate nuclear weapons.

---

[31] NPT Review and Extension Conference, 'Decisions and Resolutions Adopted by the 1995 Review and Extension Conference of the Parties to the NPT,' New York, 17 April–12 May 1995, p. 279.

[32] Hammad, Fawzy, Former Chairperson of the Egyptian Atomic Energy Authority, in a Meeting held at Cairo, on November 2003, expressed his concern that there is no timetable to obligate the NWS to implement Article VI.

2. The indefinite extension creates a blockade towards providing real change in the current rules. The threshold countries that are known to have nuclear weapons—Israel, Pakistan and India—are still non members of the NPT and it is not foreseeable that they will join it in the near future. Their goals and interests lie in keeping and developing their nuclear weapons. On the other hand, they are not willing to give up their nuclear arsenals to be parties to the treaty as NNWS, even though the international community will maintain its position in considering them undeclared NWS from the legal perspective.
3. The indefinite extension aimed at applying the concept of universality of the treaty. However, the treaty is not likely to reach this end within the current legal rules and political situation.[33]
4. After the terrorist attacks on the US destroying the twin towers of the World Trade Centre in New York in September 2001, the Bush administration—representing a State party to the NPT and one of the main supporters of the current NPT regime—threatened to use nuclear weapons against NNWS, if the US determined the need for such a measure in order to combat terrorism. Furthermore, the Nuclear Posture Review of 2001 stated that "In setting the requirements for nuclear strike capabilities, distinctions can be made among the contingencies for which the US must be prepared ... North Korea, Iraq, Iran, Syria and Libya are among the countries that could be involved in immediate, potential, or unexpected contingencies as they have had longstanding hostility towards the US and its security partners; North Korea and Iraq in particular have been chronic military concerns. All sponsor or harbour terrorists and all have active weapons of mass destruction and missile programs."[34]
5. Under the aforementioned assessment, it can be concluded that the indefinite extension of the NPT and the continuation of enforcing the current rules are still of relevance. However, after 36 years, the main goal of the treaty, to fully eliminate nuclear weapons, has not been reached. NWS did not follow the pace that NNWS have hoped for; on the contrary, nuclear arsenals are still compiling in NWS, and

[33] The case of North Korea created a unique situation within the international regime established to ensure the non-proliferation of nuclear weapons. It challenges also the concept of the universality of the NPT.

[34] NPR 2001, p. 16.

to the disappointment of NNWS, the threshold States are following the same path, i.e., India, Pakistan, and Israel.

In addition, the US, in its war against terrorism, not only threatened to use nuclear weapons against NNWS if it proved that they harbour terrorism,[35] but also decided to withdraw from the Anti Ballistic Missile Treaty,[36] which is crucial to keep the current balance of deterrence in the world and highly important to combat the proliferation of nuclear weapons and other weapons of mass destruction in general.

6. *Universality*

Providing universality to the NPT does not seem feasible as long as some States are not under its umbrella. It is hard to imagine that countries such as Israel, India or Pakistan would agree to join the Treaty as NNWS. Two countries that are parties—Iraq and North Korea—decided, at different times, to violate the treaty and make their own nuclear weapons. The Iraqi case seems to have been solved as its nuclear programme has been destroyed. The North Korean case has not yet been solved.[37] The complexity of the matter is if NWS require nuclear weapons for their security, it is impossible for them to prevent others from needing similar weapons for theirs. In the post Cold War world, it seems that countries are creating problems and tensions that did not exist under the restraints of the Cold War. NWS should exert more effort to achieve nuclear disarmament in order to encourage NNWS to remain non-nuclear. Iraq is no longer a problem after the destruction of its military capabilities and since many of its nuclear scientists have left the country. It is also important to note that Iran is cooperating with the IAEA to ensure that it is not pursuing any military capabilities and that it is only developing its nuclear infrastructure for peaceful purposes. So the real problem today lies in North Korea, India, Israel and Pakistan who have nuclear weapons and are not part of the system.

---

35 Hitchens, Theresa, 'Slipping Down the Nuclear Slope: Bush Administration Nuclear Policy Lowers Bar Against Usage,' presented at a conference on 'US Nuclear Policy and Counter Proliferation,' 26 February, 2003.

36 The US withdrew from the Treaty on 13 June 2002, in which it is stated in its third preambular paragraph that effective limits on anti-missile systems would be a "substantial factor in curbing the race in strategic offensive arms."

37 UN, 'Assuring the Success of the Non-Proliferation Treaty Extension Conference,' 1994, p. 107.

The lack of universality constitutes a threat to the NPT. This was demonstrated when India and Pakistan carried out nuclear explosions in 1998. The action alarmed the international community as it affected the whole system and questioned its validity as the sole vital regime for nuclear non-proliferation. However, reactions of the international community varied from condemning the explosions, warning States not to follow the steps of those two countries and imposing economic and military sanctions. These sanctions did not affect them as had been expected. But the fact remained that these two countries have been declared NWS. As for the ambiguous position of Israel, it has developed nuclear weapons but is following the 'bomb in the basement strategy' and so far the international community is silent towards the existence of these weapons.

These States are not members to the NPT and it is implausible that they will be in the near future, as they would have to give up their nuclear weapons to become members. This will not take place as long as these State have reason to believe that they need these weapons as a valuable deterrent. In this regard, they appear to pursue the model of the NWS who have relied on nuclear weapons as the core of their security paradigms for decades.[38]

It is clear that by an act of commission or omission, NWS have enabled more countries to acquire nuclear weapons. The international conscience has a role to play in the operation of the NPT. Defeating the universality concept and maintaining their position as non-members of the NPT assisted these countries to access the benefits of nuclear technology more than NNWS parties to the Treaty. Hence, what would the incentive be for other States to remain members of the Treaty?[39]

7. *Conclusions*

The NPT was established to prevent the proliferation of nuclear weapons. It encouraged the transfer of nuclear technology for peaceful purposes. The implementation proved that nuclear technology in the peaceful fields

[38] Nolan, Janne, 'Technology and Non-Proliferation in a Changing World Order,' Transnational Law and Contemporary Problems, Volume 2, Number 1, 1992, pp. 401–402.

[39] UN, 'Assuring the Success of the Non-Proliferation Treaty Extension Conference,' 1994, Statement made by Ambassador Ayewah, Chairman of the Third NPT Preparatory Committee of the NPT Review and Extension Conference, p. 120.

were used at its best in the developed world, however, the benefits were relatively limited in the developing countries.

On the other hand, the NPT regime was able to limit the proliferation of nuclear weapons to a great extent. However, there are still numerous failures in practice. The threat of proliferation of nuclear weapons is still at hand; more states were able to announce their development of these weapons and the world that fought for establishing the NPT has remained silent. The NPT's discriminatory nature and the failure of NWS to comply with their obligations under Article VI of the Treaty, led to the conclusion that the current regime rules are not strong enough to enforce a full non-proliferation, or to ensure the total elimination of existing nuclear weapons. Furthermore, with existing terrorist activities around the world, the threat of proliferation and using such weapons is increasing.

The NPT suffers from several inadequacies. However, these inadequacies can only be overcome if the rules undergo proper development in order to preserve the current system.

In fact, the promises that were made in the Treaty were vague. There is no time frame or specific steps agreed upon among the parties in Article VI of the NPT for the aim of realising its main goal of nuclear disarmament.

The discriminatory character of the NPT, the reluctance of NWS to meet their obligations under Article VI of the treaty and the emergence of NWS outside the NPT defeat efforts to achieve universality.

It is important to keep these issues on the international agenda. Even if it seems unlikely to be fulfilled, political and legal pressures may eventually lead to global nuclear disarmament.[40]

However, another threat today involves terrorist groups acquiring nuclear weapons. Hence, the conclusion that can be drawn here is that as long as these weapons still exist, the threat of nuclear proliferation will be persistent.

The rules of the NPT cannot be considered satisfactory either to encourage the transfer of nuclear technology for peaceful purposes or to eliminate nuclear weapons as the phrasing is too general and the obligations that are stated in the treaty do not request its member States to undertake specific measures for the implementation of such obligations.

---

40 Statement of Raouf, Tarik, Section Head, Verification and Security Policy Coordination, Office of External Relations and Policy Coordination at the IAEA, in an interview concluded at IAEA Headquarters, 2004.

Contributing States are obligated to transfer nuclear technology for peaceful purposes to Beneficiary States and to provide assistance in the field. However, there is no request to allocate funds for such assistance or criteria to identify the Contributing or Beneficiary States. Each individual State decides how to interpret these measures. Also, it is left to the credence of member States to organise means of cooperation in the field. That is one of the reasons behind the limited use of nuclear techniques around the world, especially in developing countries. Except for the assistance received through the Technical Cooperation Programmes of IAEA, developing countries would have barely had access to peaceful nuclear applications.

Article VI generated another gap in the course of its implementation. The NWS are required to pursue negotiations for the cessation of the nuclear arms race that should lead eventually to the complete elimination of nuclear weapons. However, there was no time frame mentioned in the Article to address the issue of reducing current nuclear arsenals. There were no regulations set up to ensure that NWS halt vertical proliferation by freezing the production of fissionable materials and terminating the manufacture of nuclear weapons. The article did not even set up a date for the negotiations to commence.

The NPT failure to adopt such measures weakens it. The current stagnation in the system and the lack of any will to develop rules to regulate the proper implementation of the aforementioned articles threaten the effectiveness of the treaty and may encourage States to reassess their position towards the treaty, as did North Korea.

There is a need to amend the NPT to close the loop-holes inherent in its rules and consequently to resolve the inadequacies in the system. These amendments may be adopted in the form of a protocol to be annexed to the NPT. This protocol should amend Articles III, IV ad VI as follows:

- Providing NNWS with assurances relating to their access to nuclear material and technology, through instigating a system of cooperation between Supplier and Recipient States. This system should work in transparency and should include members to represent the interests of both groups of States (Contributing or Supplier States on one side and on the other, Beneficiary or Recipient States).
- Supplier States should cooperate with IAEA and inform it of their nuclear transfers to NNWS in order to facilitate the implementation of safeguards.

- Adopting efficient measures to ensure the adherence of the NWS to their obligations under the treaty. The first of these should be the adoption of a time table to reduce their nuclear arsenals and measures to halt any developments of new types of nuclear explosive devices.

*

SECTION TWO
TRANSFER OF NUCLEAR TECHNOLOGY FOR PEACEFUL PURPOSES UNDER IAEA RULES

1. *Introduction*

The IAEA was established on 29 July 1957 to promote peaceful uses of nuclear energy for the benefit of humanity. One of the objectives of the IAEA is to ensure that nuclear projects, safeguarded by bilateral or multilateral arrangement, maintain their peaceful nature. Since 1970, after the signing of NPT, the IAEA has been entrusted with verifying the compliance of its member States to the treaty obligations.

As stipulated in Article II of the IAEA Statute, the Agency has dual objectives to "seek to accelerate and enlarge the contribution of atomic energy to peace, health and prosperity throughout the world" and to "ensure, so far as it is able, that assistance provided by it … is not used in such a way as to further any military purpose."

To encourage the development of peaceful applications of nuclear energy (Article IV of NPT), IAEA provides technical assistance directly to member States through Agency projects, or acts as an intermediary in supplying assistance. To assure that any nuclear fuel or technology it provides is not used to further military purposes, IAEA designs and administers safeguards on Agency projects in conjunction with regional nuclear- monitoring organisations.

Article III of the NPT requires each NNWS to accept safeguards as set forth in an agreement with IAEA, "on all special fissile material whether it is being produced, processed or used in all peaceful nuclear activities within the territory" of a State party. However, this obligation is only imposed on NNWS. This is considered a real weakness in the NPT rules which means that safeguards do not extend to the NWS.[41]

---

[41] UN, 'Assuring the Success of the Non-Proliferation Treaty Extension Conference,' 1994, p. 97.

NPT parties negotiated with IAEA a model NPT safeguards agreement that was approved by the IAEA Board of Governors and has become the basis for negotiation of all bilateral NPT safeguards agreements between IAEA and NPT NNWS.[42]

In this section, we are going to discuss the role of IAEA in international cooperation to promote peaceful uses of nuclear energy and some concerns related to it. Discussing IAEA legal authority under NPT safeguards agreements, the international support that it enjoys and developments to strengthen IAEA safeguards to ensure its effectiveness shall follow.

2. *Role of* IAEA *in International Cooperation*

The IAEA plays an important role in technical cooperation, especially in assisting developing countries with peaceful uses of nuclear energy through the development of effective programmes aimed at improving their scientific, technological and regulatory capabilities. (Chapter 1 described the fields of nuclear techniques implemented in member States through IAEA Technical Cooperation Programme). The NPT 2000 Review Conference emphasised that strengthening IAEA safeguards should not adversely affect the resources available for technical assistance and cooperation.[43]

IAEA's role is expanding in the field of nuclear information exchange. It organises conferences and symposia for this matter. It has a computer-based International Nuclear Information System to compile and disseminate information concerning peaceful uses of nuclear energy.[44]

---

42 Rockwood, Laura, 'Legal Instruments Related to the Application of the Safeguards,' 1998–2000.

43 NPT Review Conference, NPT/CONF.2000/28, 'Final Document of the 2000 Review Conference of the Parties to the NPT,' 24 May 2000, p. 6.

44 Shaker, Mohamed, 'The Nuclear Non-Proliferation Treaty: Origin and Implementation 1959–1979,' 1980, p. 361.

3. IAEA's Member States Legal Authority under Safeguards Agreements

3.1. *Under Safeguards Agreement outside NPT Framework*

The first Safeguards Document (INFCIRC/26) was worked out by interested Governments and the IAEA Secretariat in 1959 and 1960 and was approved by the BG on 31 January 1961. It comprised principles and procedures for the application of safeguards to small reactors. This document was extended to larger reactors by decision of the Board on 26 February 1964. In 1964 and 1965, a completely revised Safeguards Document was worked out by a group of government experts and approved by the Board after unanimous concurrence by the General Conference in September 1965 (INFCIRC/66). Annex I to INFCIRC/66, which includes provisions for reprocessing plants, was approved by the Board in 1966, and Annex II, which includes provisions for safeguarded nuclear material in conversion and fuel fabrication plants, was adopted by the Board in 1968. With its two annexes, the Safeguards Document is now referred to as INFCIRC/66/Rev.2. Its provisions are incorporated by reference in Safeguards Agreements.[45]

The principal purpose of INFCIRC/66 is to establish a system of control enabling the Agency to comply with this statutory obligation with respect to activities of Member States in the field of peaceful uses of nuclear energy, as provided by Article III.A.5 of the IAEA Statute. The said article authorises the Agency to "establish and administer safeguards designed to ensure that special fissionable and other materials, services, equipment, facilities, and information made available by the Agency or at its request or under its supervision or control are not used in such a way as to further any military purpose". This Article further authorises the Agency to "apply safeguards, at the request of the parties, to any bilateral or multilateral arrangement, or at the request of a State, to any of that State's activities in the field of atomic energy". Article XII.A sets forth the rights and responsibilities that the Agency is to have, to the extent relevant, with respect to any project or arrangement which it is to safeguard.[46]

---

45 Rockwood, Laura, 'Legal Instruments Related to the Application of the Safeguards,' 1998–2000.

46 IAEA, 'The Agency's Safeguards System (1965, as Provisionally Extended in 1966 and 1968),' INFCIRC/66/Rev.2, General Distribution, 16 September 1968.

In June 1961, the BG adopted a document referred to as the Inspectors Document (GC (V)/INF/39, Annex). It covers four different areas of inspection activities, including designation of Agency inspectors, notification of inspections, conduct of inspection and rights of access and privileges and immunities of inspectors. This document is also incorporated in and only relevant to INFCIRC/66-type agreements.[47]

The INFCIRC/66 system was designed chiefly to define the safeguards to be applied to individual plants and shipments of fuel, although it could and in certain cases did cover all nuclear trade between two Member States and/or all nuclear activities in a State. The leading industrial NNWS considered the system too flexible. In their view, INFCIRC/66 left too many decisions to the discretion of IAEA Secretariat and was too liberal in setting up standards for the frequency of IAEA inspections.[48]

Safeguards agreements concluded in accordance with INFCIRC/66/Rev.2 incorporate the Agency's statutory right of access to all persons, places and information relevant to the implementation of safeguards. They prohibit the use of safeguarded items in such a way as to further any military purpose, which includes a proscription against their use of any nuclear explosive device or any other military purpose (including nuclear naval propulsion). According to these agreements, the State undertakes not to use safeguarded items for "any military purposes". In 1974, the IAEA Director General (DG) proposed, and the Board accepted, an interpretation of that undertaking as precluding the use of safeguarded items of any nuclear explosive device, whether intended for peaceful or non-peaceful ends, owing to technical impossibility of distinguishing between a nuclear explosive device for peaceful uses and one for military uses. Although a small number of States expressed reservations about this interpretation, all INFCIRC/66/Rev.2 safeguards agreements since 1975 have incorporated a basic undertaking which expressly precludes the use of safeguarded items for the manufacture of any nuclear weapon or to further any other military purpose or for the manufacture of any other nuclear explosive device. The scope of INFCIRC/66-type agreements over the years has been expanded with the approval of the Board to include provisions for safeguarding such

---

[47] Rockwood, Laura, 'Legal Instruments Related to the Application of the Safeguards,' 1998–2000.

[48] Fischer, David, 'Safeguards, Past, Present and Future,' IAEA Bulletin, Volume 394, October 1997.

items as non-nuclear materials (such as heavy water and zircaloy), non-nuclear facilities (heavy water production plants), and transferred technology.[49]

IAEA Obligations according to INFCIRC/66[50]

1. While concluding safeguards, the Agency shall avoid hampering a State's economic or technical development and shall be consistent with prudent management practices required for the safe conduct of nuclear activities. This shall be applied through consultations between the State or States and the DG of the IAEA.
2. In no case shall the Agency request a State to stop the construction or operation of any *principal nuclear facility*, to which safeguards procedures of the Agency extend, except by explicit decision of the Board.
3. In implementing safeguards, the Agency shall take every precaution to protect commercial and industrial secrets. The staff of the Agency shall not disclose, except to the DG and to such other members of the staff as the DG may authorise to have such information by reason of their official duties, any commercial or industrial secret or any other confidential information coming to their knowledge by reason of the implementation of safeguards by the Agency.
4. The Agency shall not publish or communicate to any State, organisation or person any information obtained by it in connection with implementing safeguards, with the following exceptions:
   a. Specific information relating to such implementation in a State may be given to the Board and to such Agency staff members which require such knowledge by reason of their official duties in connection with safeguards, but only to the extent necessary for the Agency to fulfil its safeguards responsibilities;
   b. Summarised lists of items being safeguarded by the Agency may be published upon a decision of the Board; and
   c. Additional information may be published upon decision of the Board if the directly concerned States agree.

[49] Rockwood, Laura, 'Legal Instruments Related to the Application of the Safeguards,' 1998–2000.

[50] IAEA, INFCIRC/66/Rev.2, 'The Agency's Safeguards System (1965, as Provisionally Extended in 1966 and 1968),' General Distribution, 16 September 1968.

When Shall the Agency Implement Safeguards in a State?[51]

The Agency shall implement safeguards in a State if:

1. The Agency has concluded with the State a *project agreement* under which materials, services, equipment, facilities or information are supplied, and such agreement provides for the application of safeguards; or
2. The State is a party to a bilateral or multilateral arrangement under which materials, services, equipment, facilities or information are supplied or otherwise transferred, and:
   a. All Parties to the arrangement have requested the Agency to administer safeguards; and
   b. The Agency has concluded the necessary *safeguards agreement* with the concerned State; or
3. The Agency has been requested by a State to safeguard certain nuclear activities under its jurisdiction, and the Agency has concluded the necessary *safeguards agreement* with such State.

Which Nuclear Materials Are Subject to Safeguards under INFCIRC/66?[52]

1. *Nuclear material* shall be subject to the Agency's safeguards if it is being or has been:
   a. Supplied under a *project agreement*; or
   b. Submitted to safeguards under a *safeguards agreement* by the parties to a bilateral or multilateral arrangement; or
   c. *Unilaterally submitted* to safeguards under a *safeguards agreement*; or
   d. Produced, processed, or used in a *principal nuclear facility* which has been:
      i. Supplied wholly or substantially under a *project agreement*; or
      ii. Submitted to safeguards under a *safeguards agreement* by the parties to a bilateral or multilateral arrangement; or
      iii. *Unilaterally submitted* to safeguards under a *safeguards agreement*; or

[51] Ibid.
[52] Ibid.

e. Produced in or by use of safeguarded *nuclear material*; or
f. Substituted for safeguarded nuclear material, pursuant to paragraph 26(d), which stipulates that the State or States concerned have agreed with IAEA to place under safeguards, as a substitute, such amounts of the same element, not otherwise subject to safeguards, as the Agency has determined to contain fissionable isotopes:
    i. Whose weight (with due allowance for processing losses) is equal to or greater than the weight of fissionable isotopes of material with respect to which safeguards are to terminate; and
    ii. Whose ratio by weight to the total substituted element is similar to or greater than the ratio by weight of the fissionable isotopes of the material with respect to which safeguards are to terminate to the total weight of such material; provided that the Agency agrees to the substitution of plutonium for uranium-235 contained in uranium whose *enrichment* is not greater than 0.05 (5.0%).

2. A *principal nuclear facility* shall be considered as substantially supplied under a *project agreement* if the Board has so determined.

Duration and Termination of INFCIRC/66 Agreements

INFCIRC/66/Rev.2 makes reference to the "desirability" of providing for the continuation of safeguards with respect to produced special fissionable material and to any materials substituted there for. In 1973, the Board expressed concern about the need for safeguarding such material after the expiry of a safeguards agreement. As a consequence, since 1974, the duration of 66-type agreements has been tied to the actual use in the recipient State of supplied material or items, rather than to fixed periods of time. Under these agreements, safeguards are required to continue on all safeguarded items, including subsequent generations of produced nuclear material derived from safeguarded material or facilities, until safeguards are terminated in accordance with the revisions of INFCIRC/66/Rev.2.[53]

[53] Ibid.

3.2. *Under NPT Safeguards Agreements*

NPT parties are requested to conclude agreements with IAEA to determine the scope of its legal authority to inspect sites in member States. The IAEA has adopted a narrow interpretation as a result of tensions between its role in verifying compliance with the NPT and its sensitivity to States' concerns about the infringement of their sovereignty. Internal constraints on safeguard activities prevented the IAEA from invoking its special inspections procedures at undeclared sites, which in turn created a gap in the non-proliferation regime that the Iraqi clandestine nuclear weapons programme was able to exploit.

In April 1971, IAEA Board approved a model NPT Safeguards Agreement known as INFCIRC/153.

Legal Effect of INFCIRC/153 on States' Safeguards Obligations under the NPT

Introduction
According to Article III of the NPT, NNWS undertake to accept IAEA safeguards "on all source or special fissionable material in all peaceful nuclear activities within its territory, under its jurisdiction or carried out under its control anywhere, for the exclusive purpose of the verifying that such material is not diverted to nuclear weapons or other nuclear explosive devices."

Inspections under INFCIRC/153
INFCIRC/153 provides for IAEA to conduct three types of inspections: routine inspections, ad hoc inspections and special inspections. INFCIRC/153 included inspections provisions to confirm, *inter alia*, "that reports are consistent with records" and with "the location, identity, quantity and composition of all nuclear material subject to safeguards under the Agreement." Routine inspections cannot commence until the State and IAEA have agreed on "Subsidiary Arrangements." Subsidiary Arrangements are detailed agreements between the State and the IAEA that specify, for each facility, subject to routine inspections, "how the procedures laid down in the Agreement are to be applied." The purposes of ad hoc inspections include confirmation of information contained in the initial report and the verification of inventory changes after the initial report and prior to the commencement of routine inspections. As for special inspections, the Agency may request them if it considers "that

information made available by the State, including explanations from the State and information obtained from routine inspections, is not adequate for the Agency to fulfil its responsibilities under the Agreement." The chief responsibilities referred to here are responsibilities of the Agency to verify that the State has not concealed undeclared nuclear material or diverted declared nuclear material for nuclear explosive purposes. During special inspections, the IAEA may obtain access, in agreement with the state, to locations other than those declared by the latter to contain nuclear material. Any disagreement concerning the need for additional access is to be resolved by referring the dispute to the Board, and then, if necessary, to an arbitral tribunal. If the Board determines that the need for access to an undeclared location is "essential and urgent," then the Board may bypass the arbitral tribunal and "call upon" the state to grant access "without delay." (Paragraph 20–22 of INFCIRC/153)[54]

Former DG Blix confirms that if IAEA demands a special inspection of an undeclared site, a refusal would violate safeguards agreement.[55] Hence, the IAEA can resort to the SC to take action against the non-compliant State.

The 2000 Review Conference reaffirmed that IAEA is the competent authority responsible for verifying and assuring compliance with its safeguards agreements with States parties, with a view to preventing diversion of nuclear energy from peaceful uses to nuclear weapons or other nuclear explosive devices.[56]

Constraints on the IAEA's Legal Authority

The 2000 Review Conference noted the considerable increase in the safeguards responsibilities of the Agency since 1995 and the financial constraints under which the IAEA safeguards system is functioning. It called upon all States party to the treaty to continue their political, technical and financial support to IAEA in order to ensure that the Agency is able to meet its safeguards responsibilities, especially with growing needs to develop new safeguards techniques to deal with uranium enrichment and other nuclear installations being decommissioned.[57]

---

54 Sloss, David, 'It's not broken, so don't fix it: the IAEA safeguards system and the NPT,' Virginia Journal of International Law, Volume 35, Issue 4, 1995, p. 856.

55 Ibid, p. 856.

56 NPT Review Conference, NPT/CONF.2000/28, 'Final Document of the 2000 Review Conference of the Parties to the NPT,' 24 May 2000, p. 3.

57 Ibid, p. 6.

The risk of timely detection of diversion of nuclear material from peaceful activities to nuclear explosive devices is not an easy task, considering that the application of safeguards is becoming more demanding and the constraints on the budget of the IAEA.[58]

Right of Access to Information

The IAEA has taken several steps to address problems related to access to triggering information. For example, in February 1992, the BG endorsed a proposal for States to provide design information to the IAEA at the time they decide to construct a new nuclear facility or modify an existing facility. It has also acknowledged the right of the Agency to receive information from additional sources, which could include IAEA inspectors, governments, import and export records for nuclear material and sensitive technology and news accounts. In February 1993, the IAEA Board also "endorsed a reporting scheme on imports and exports of nuclear material and exports of specified equipment and non-nuclear material." Most recently, in June 1995, the Board approved additional measures to strengthen the safeguards system. The most significant of these measures, for purposes of gaining access to triggering information, is a plan for the IAEA to make greater use of environmental monitoring techniques to provide early warning of clandestine nuclear activities.[59] These steps played an important role in uncovering the Iraqi clandestine nuclear programme.

However, the IAEA will remain dependent on member States to gain access to the triggering information it needs. This, however, constitutes a problem when a State is incompliant with its safeguards obligations. It leads to prolonging the procedures and implementation of measures undertaken by IAEA to correct the situation in the State concerned. There should be rules to request that supplier States inform the Agency of their nuclear exports as a means to close the loophole in the rules as they stand.

Right of Access to Sites

Under routine inspection procedures, IAEA inspectors have the right to access sites stated in an annual report presented by an NNWS about its nuclear activities. The scope, limit and time of inspections are concluded

[58] Goldblat, Josef, 'Twenty Years of the NPT implementation and prospects,' 1990, p. 17.

[59] Sloss, David, 'It's Not Broken, so Don't Fix It: the IAEA Safeguards System and the NPT,' 1995, pp. 888–889.

after consultations between IAEA and NNWS. The inspections' main goal is to ensure the non-diversion of nuclear material and equipment from peaceful to military purposes.

The IAEA has the right to conduct different kinds of inspections; according to the IAEA Statute, the Agency has the right to access sites under special inspection procedures in order to gain "access to information or locations in addition to the access specified" for routine and ad hoc inspections.[60]

However, the right of the IAEA to access undeclared sites under INFCIRC/153 is always restricted either by consent of the State concerned or by the BGS' consent. IAEA's responsibility is to ensure the compliance of NNWS to the NPT, which should be implemented in accordance with procedures required in safeguards agreement with each individual State. This restriction can not be overruled because it is the legal basis on which the IAEA conducts its inspections in member States. Hence, even if IAEA experts attempt to grant the NPT safeguards Articles a wide or a narrow interpretation, the balance between IAEA rights and the State's sovereignty has to be respected. This has been and will remain the proper approach to deal with safeguards rules. IAEA has discovered that INFCIRC/153 was not sufficient to provide it with further powers over declared and undeclared nuclear facilities. The INFCIRC/153 does not permit the Agency free access to all sites anywhere in a country. To acquire a legal basis for more access rights in an NNWS, the IAEA developed the Additional Model Protocol to IAEA Safeguards Agreement (INFCIRC/540).

Before requesting that a State submits to a special inspection, the DG of the IAEA must determine that reasonable grounds exist for suspecting illegal nuclear activities. As former DG Blix asserts, States complying with their NPT obligations would likely grant access to IAEA inspectors to dispel groundless suspicions. By the same token, any state violating the NPT would almost certainly deny inspectors access to suspect facilities outside routine inspection areas thereby giving rise to suspicions that the State is seeking to conceal a violation. The IAEA would be unable either to verify an uncooperative State's compliance with the NPT or to prove a violation of the treaty. Still, the IAEA does not have to confirm a violation

---

60 Bunn, George, 'Inspection for Clandestine Nuclear Activities: Does the NPT Provide Legal Authority for the IAEA's Proposals for Reform?', Nuclear Law Bulletin, Bulletin 57, June 1996, p. 21.

to find an NPT state incompliant with its NPT safeguards agreement. The DG must report the denial of the inspection request to the BG, which may use its emergency procedures to call on the inspected State to comply with the request. If, after reviewing the DG's report, the Board cannot "verify that there has been no diversion of nuclear material required to be safeguarded under the Agreements to nuclear weapons," the IAEA may resort to IAEA statutory procedures for situations of non-compliance. First, the BG may impose sanctions delineated in Article XII.C. Second, it may report the non-compliance to the UN SC and GA,[61] in accordance with the Charter of the United Nations, in upholding compliance with IAEA safeguards agreements and ensuring compliance with safeguards obligations.[62]

The authority of the IAEA to conduct special inspections for detecting covert nuclear weapons development is ambiguous. In Iraq, the UN cease-fire agreement and subsequent SC resolutions granted the IAEA far more extensive authority. The SC authorised inspectors not only to verify Iraq's compliance with its NPT obligations, but also to prevent Iraqi misuse of nuclear material for military purposes in the future. This extraordinarily broad power also included the authority to search out and supervise the destruction of suspect sites in conjunction with the UN Special Commission on Iraq (UNSCOM), and is far more extensive than that which the Agency exercises under the NPT safeguards system. Thus, the extraordinarily intrusive inspections in Iraq are unique and do not necessarily indicate that the Agency has the authority to verify the absence of nuclear weapons programmes at undeclared sites under the NPT. So, if the IAEA is to detect undeclared nuclear facilities, it cannot be constrained by the traditional narrow interpretation of its NPT special inspections authority. While States could amend the NPT or the IAEA Statute to grant the IAEA authority to conduct special inspections, this would be both difficult and unnecessary. Instead, constructing the provisions of NPT safeguards agreements, in light of the verification role of the IAEA under the NPT, should provide the Agency with sufficient authority to conduct special inspections at undeclared sites.[63]

---

[61] NPT Review Conference, NPT/CONF.2000/28, 'Final Document of the 2000 Review Conference of the Parties to the NPT,' 24 May 2000, p. 3.

[62] Ibid.

[63] Sloss, David, 'It's Not Broken, so Don't Fix It: the IAEA Safeguards System and the NPT,' 1995, p. 858.

4. *Developments to Strengthen IAEA Safeguards*

Since the discovery of the Iraqi clandestine nuclear programme in 1991, the IAEA has taken several steps to enhance the NPT safeguards system. This approach by the IAEA was reaffirmed in the NPT 2000 Review Conference, whereby States were encouraged to support and implement the BG decisions aimed at further strengthening the effectiveness and efficiency of IAEA safeguards.[64]

4.1. *Additional Measures to Strengthen the Special Inspections Procedures*

Existing IAEA/NPT safeguards agreements allow the IAEA to develop its techniques to strengthen its special inspection procedures as long as it does not impose new legal obligations on member States without their prior consent.

The 2000 Review Conference noted the high priority that IAEA attaches, in the context of furthering the development of strengthened safeguards systems, to integrating traditional nuclear-material verification activities with new strengthening measures looking forward to an expeditious conclusion of this work. It has also recognised that measures to strengthen the effectiveness and improve the efficiency of the safeguards system, with a view of providing credible assurance of non-diversion of nuclear material from declared activities and of the absence of undeclared nuclear material and activities, must be implemented by all States party to the NPT, including NWS.[65]

Granting access to the IAEA or expanding its rights of access to information is not beneficial to NNWS. There are no foreseeable security incentives to States if they accept the expansion of the IAEA role. Based on the Iranian experience, which will be detailed later, cooperation and expansion of IAEA rights to access nuclear sites is among the reasons behind Iran's diminished ability to develop its nuclear capabilities for peaceful purposes. The attempt to block its efforts to develop its enrichment technology and centrifuge production constitutes in and of itself a violation of the NPT, which grants NNWS inherent rights to develop nuclear

[64] NPT Review Conference, NPT/CONF.2000/28, 'Final Document of the 2000 Review Conference of the Parties to the NPT,' 24 May 2000, p. 4.

[65] Ibid, p. 5.

capabilities for peaceful purposes. Hence, even if an NNWS grants the IAEA full access to its sites, this is not a guarantee that it will not hamper the state's security or interests.

4.2. *The IAEA's "Programme 93 + 2"*

At the beginning of 1993, the IAEA Secretariat prepared a comprehensive plan called "Programme 93 + 2" intending to seek Board approval of the plan in 1995. At meetings in March and June 1995, the BG endorsed the general direction of the efforts of the Secretariat to assure that future declarations of NNWS to the IAEA about their nuclear activities are not only correct as far as they go, but are complete in that there are no undeclared, clandestine nuclear activities. The Board also took note that many of the new measures proposed were within the existing legal authority of the IAEA as expressed in safeguards agreements with NNWS. The new measures had two dimensions; Measures of "Part 1" include seeking additional information than that provided in the declaration of the State that are directly implemented. "Part 2" measures are those that would require "complementary" authority according to the Secretariat; that is, they need new legal instruments to make them obligatory for States. This came in 1997 in the form of the Model Additional Protocol to safeguards agreements (INFCIRC/540).[66]

The "93 + 2" plan also calls for access to more sites than those allowed by the existing safeguards agreements; for example, to locations beyond "strategic points" in the nuclear fuel cycle now subject to routine inspections. Existing safeguards agreements with the IAEA permit inspection beyond "strategic points" before the running of a facility in order to check the design provided by the inspected nuclear organisation against the actual facility as built. For ad hoc inspections, going beyond strategic points is also now permitted. Where that is so, access would be a "Part 1" measure. For routine inspection of non-strategic points in an operating nuclear facility, however, "Part 2" approval would be required—except where authorised under the provisions for special inspections.[67]

Inspection of nuclear-related activity sites not previously declared (but to be declared pursuant to Part 2) would follow as a result of the new

---

[66] Bunn, George, 'Inspection for Clandestine Nuclear Activities: Does the NPT Provide Legal Authority for the IAEA's Proposals for Reform?', 1996, p. 11.

[67] Ibid, p. 12.

legal obligation to provide additional information in a revised declaration. Beyond this, there is the question of whether IAEA inspectors may permissibly search for nuclear weapons activities at sites where no nuclear-related activity or nuclear material has been reported. Another new measure is unannounced (no-notice) inspections. These are currently permitted at strategic points, though they do not occur often. No-notice inspections at other points and at locations without strategic points would, of course, require Part 2 approval.[68]

The language of Article III of the NPT establishes broad standards and purposes for safeguards that NNWS members must accept in an agreement with the IAEA. In the context of 93 + 2 measures, it raises the following questions:

i. What standards are contemplated by "in accordance with the Statute of the IAEA and the Agency's safeguards system"?
ii. What "obligations assumed under this treaty" is it the purpose of safeguards to verify?
iii. What are the limitations imposed by the NPT language on IAEA inspection?

What Standards Are Contemplated by "in accordance with the Statute of the IAEA and the Agency's" in Article III.1 of the NPT?

The Statute of the IAEA in Article III.5 authorises the Agency to establish safeguards "to ensure that special fissionable material and other materials, services, equipment, facilities and information … are not used in such a way as to further any military purpose; and to apply safeguards, at the request of the parties, to any bilateral or multilateral arrangement, or at the request of a State, to any of that State's activities in the field of atomic energy." This is the safeguards system of the Agency before the NPT (INFCIRC/66). IAEA applies safeguards upon the request of a State, so the implementation was established on a case by case basis. However, INFCIRC/153 is applied to all nuclear material and equipment in a NNWS member to the NPT (Article III of the NPT). The scope of safeguards in a NNWS territory is governed by the provisions of its safeguards agreement with the Agency.

[68] Ibid.

What "Obligations Assumed under this Treaty ..." (in the Language of Article III.1) is it the Purpose of Safeguards to Verify?

Inspection should cover early steps that may lead to the production of explosive nuclear material in, for example, plants for plutonium separation, uranium enrichment or fuel fabrication, as well as in reactors and nuclear-material storage facilities, all for the purpose of assuring that plutonium or enriched uranium is not diverted to nuclear explosives. INFCIRC/66 and 153 both include provisions for inspection of reactors and other fuel cycle facilities. Moreover, the scope of Article II of the NPT obligation, not to "manufacture" nuclear explosives, is broad enough to include not only these facilities but steps before the final assembly of a nuclear explosive device. Therefore, the IAEA is authorised by Article III.1 to inquire about and inspect those steps.[69] This obligation goes beyond manufacturing to include the obligation of a NNWS "not to acquire" a nuclear explosive device.

However, the obligation to inspect early steps (including plutonium separation or uranium enrichment) is intended to ensure the non-diversion of the materials in question into nuclear explosives. This should be implemented in its proper context—not used to hamper the right of NNWS to acquire and produce such material for peaceful purposes and scientific research.

What Limitations Does the NPT Language Impose on IAEA Inspection?

Limitation of "Peaceful" Nuclear Activities

NPT relies on IAEA to conduct inspections to ensure the non-diversion of nuclear material and equipment to military purposes. Hence, the language of Articles II and III of the NPT binds the work of the IAEA inspectors. This was highlighted in INFCIRC/153. Inspection should take place in sites that are connected to nuclear activities in a NNWS. As a result, NNWS should not engage in activities to obstruct the work of the inspectors as long as it is aimed at guaranteeing the peaceful purposes of nuclear activities in the State concerned.

---

[69] Ibid, p. 17.

Limitation of Activities Where Nuclear Material Is Present

In practice, IAEA focuses its inspections on nuclear facilities where nuclear material is used. Unless it has substantial evidence to invoke either special or ad hoc inspections, the Agency follows this interpretation of its Statute and NPT while applying its safeguards. Therefore, INFCIRC/153 focused on reports, accounting systems and locations where nuclear material and equipment exist in a member State. This was clear in the case of Iran, where the main discrepancies between Iran and the Agency related to the accounting system and the failure of Iran to report to the Agency, on a timely basis, the existence of certain nuclear material and the production of certain equipment that can be used in enriching uranium. This constituted a breach of Iran's Safeguards Agreement with the IAEA and Iran was requested to undertake all necessary corrective measures in this regard.

Limitation of Nuclear Activities that Have Been Declared and are Located at "Strategic Points"

In the case of "strategic points," the preamble of the NPT supports "application of the principle of safeguarding effectively the flow of source and special fissionable materials by use of instruments and other techniques at certain strategic points." This principle had a major impact on the negotiation of the NPT model safeguards agreement. INFCIRC/153 reflects that impact by prohibiting inspectors, during routine inspections, from access to nuclear activities except at strategic points. However, by its terms, the language of the preamble is a statement of objective; it does not limit safeguards to strategic points at all times and in all circumstances. Indeed, INFCIRC/153 designated cases in which the principle is not to be applied. Thus, INFCIRC/153 constitutes a negotiated application of the principle, and an interpretation of Article III showing that its language does not require safeguards to be applied only at strategic points. Indeed, the authority to inspect undeclared activities would disappear if strategic points designated in declared nuclear activities were all that could be inspected. Safeguards agreements can thus be supplemented consistent with Article III to permit inspection at other non-strategic locations besides those permitted by existing safeguards agreements.[70]

70 Ibid, pp. 21–22.

5. *Conclusions*

In concluding this brief review of the activities of the IAEA, it must be said that much is and will be expected from the Agency in the promotion of peaceful uses of nuclear energy as a result of the NPT. Although the results so far achieved in this respect are significant, they still fall short of the high expectations of the developing world. Initial steps have been taken for securing more technical assistance funds and financing the introduction of nuclear power plants in developing countries. Better conditions for the procurement of special fissionable materials have been promised. In the field of information exchange, difficulties still persist as a result of commercial competition and industrial secrets.

Under NPT safeguards agreements, the IAEA has legal authority to verify the absence of covert nuclear weapons programmes. The failure of the IAEA to safeguard actively hidden nuclear weapons programmes is attributed to non-legal constraints that have traditionally hampered the attempts of the Agency to fully exercise its authority under the NPT. The discovery of the Iraqi covert nuclear weapons programme made it clear that the IAEA would have to expand the exercise of its authority if it is to verify the compliance of States with their obligations.

It should be noted that the NPT clause setting forth the safeguards requirement has been applied in a way that has at times benefited non-parties more than parties. Whereas parties are subject to NPT safeguards covering all their peaceful nuclear activities, nuclear activities of non-parties are partially safeguarded. Safeguards in these States apply exclusively to imported items—individual installations or material—while part of the nuclear fuel cycle may remain unsafeguarded.[71] India and Iran are good examples for comparison: India benefited from being a non-NPT party and developed its nuclear capabilities as much as it deemed necessary, given its scientific and security needs. However, Iran is facing problems in developing its nuclear infrastructure for peaceful purposes, even though it is still committed to its obligations under the NPT and continues to cooperate with the IAEA to ensure full compliance with its safeguards obligations.

The rules of NPT/IAEA safeguards are not sufficient as they stand to ensure full compliance by NNWS to their safeguards obligations. Although the IAEA has taken major steps to develop its inspection techniques, it

[71] Ibid, pp. 16–17.

relies massively on the cooperation of NNWS to provide it with adequate information about its nuclear activities. The rules do not place obligations on Supplier States. If a NNWS does not inform the Agency about certain activities, the latter has no means of finding out by itself. It requires the cooperation of other member States in this regard. There is no international instrument and it is absent from discussions on nuclear safeguards. The IAEA should activate the arbitral procedures, stipulated in INFCIRC/153, so as not to play the role of a plaintiff and a judge simultaneously. Paragraph 22 of INFCIRC/153 provides impartial procedures to judge the compliance of a State with its safeguards agreement. It stipulates that "The Agreement should provide that any dispute arising out of the interpretation or application thereof except a dispute with regard to a finding by the Board under paragraph 19 above or an action taken by the Board pursuant to such a finding, which is not settled by negotiation or another procedure agreed to by the parties should, on the request of either party, be submitted to an arbitral tribunal composed of the following: each party would designate one arbitrator, and the two arbitrators so designated would elect a third, who would be the Chairman. If, within 30 days of the request for arbitration, either party has not designated an arbitrator, either party to the dispute may request the President of the ICJ to appoint an arbitrator. The same procedure would apply if, within 30 days of the designation or appointment of the second arbitrator, the third arbitrator had not been elected. A majority of the members of the arbitral tribunal would constitute a quorum, and all decisions would require the concurrence of two arbitrators. The tribunal would fix the arbitral procedure. The decisions of the tribunal would be binding on both parties." Even when there is a dispute arising between the Agency and a member State regarding the interpretation or the application of a safeguards agreement, the BG does not instigate these procedures and prefers to use the stick by threatening to transfer the matter to the SC. Although, both are within the powers of the Board, yet it might nevertheless improve the system and ensure transparency and impartiality to commence using this dispute settlement mechanism.

*

SECTION THREE
INTERNATIONAL COOPERATION TO TRANSFER NUCLEAR TECHNOLOGY FOR PEACEFUL PURPOSES

1. *Introduction*

International efforts to promote the use of nuclear energy are bound by the monopoly of suppliers and the financial ability of recipients. Very few States possess the required technical know-how and adequate financial resources for the new technology, therefore international cooperation is important to provide recipient States with nuclear technology.[72]

Promoting the use of atomic energy is not only bilateral. It is mainly achieved through world-wide multilateral arrangements drawn up by the IAEA, EURATOM, the NEA within the Organisation for Economic Cooperation and Development (OECD), CERN in Geneva, African Regional Co-operative Agreement for Research, Development and Training Related to Nuclear Science and Technology (AFRA), and Organisation of Nuclear Energy Producers (OPEN).[73]

International cooperation in nuclear R&D is needed to enhance overall efficiency of national efforts and to facilitate technological development. Governments and industries would benefit from pooling resources and carrying out studies jointly rather than separately. As national nuclear R&D budgets are shrinking, co-ordinated strategies for investments in capital-intensive R&D equipment would facilitate technological progress and safety enhancement. Given their experience in cooperation and joint projects, international organisations like NEA can play an important role in this regard. One of the challenges of international cooperation in a competitive environment, for nuclear energy as for other advanced technologies, will be to integrate the work of companies and business associations into governmental efforts.

In this section, the following shall be discussed:

- Channels of cooperation, bilateral and regional, in the field of transfer of nuclear technology for peaceful purposes;
- Rules of US as an example of a major Supplier State and its national rules around exporting nuclear technology to NNWS.

---

[72] Pelzer, Dr. Norbert, 'The Nature and Scope of International Cooperation in Connection with the Peaceful Uses of Atomic Energy, and its Limits—an Assessment,' Nuclear Law Bulletin, Bulletin 27, June 1981, p. 36.

[73] Ibid.

2. *Channels of Cooperation*

Article IV of the NPT stipulated the right of parties "to contribute alone or in cooperation with other States" to the further development of peaceful applications for nuclear energy.

Channels of cooperation did not receive enough attention during NPT negotiations. At the Conference of NNWS of 1968 they were, however, at the centre of debates and were the subject of many resolutions adopted by the Conference on the peaceful application of nuclear energy.[74]

The Review Conferences of the NPT, in accordance with Article IV of the Treaty, reaffirmed the undertaking of supplier States in contributing alone, or together with other States or international organisations, to further developing applications of nuclear energy for peaceful purposes, especially in the territories of NNWS party to the treaty, with due consideration for the needs of the developing areas of the world.[75]

2.1. *Bilateral Cooperation*

Bilateral agreements are concluded between States to promote the use of nuclear energy. Their scope ranges from supplying uranium, to supplying nuclear installations. The main suppliers are European States, along with US, Canada and the former USSR. There are also agreements with very specific objectives, e.g. research into reactor safety, developing fast breeder reactors, or prospecting and exploring for uranium ore.[76]

Bilateral agreements have been, and remain, the main channel for providing fissionable materials, and not the IAEA. They have also been the only channel through which developing countries have financed the construction of nuclear power stations.[77] Unfortunately, there are no statistics or surveys available to ensure the validity of this statement.

---

[74] Shaker, Mohamed, 'The Nuclear Non-Proliferation Treaty: Origin and Implementation 1959–1979,' 1980, p. 338.

[75] NPT Review Conference, NPT/CONF.2000/28, 'Final Document of the 2000 Review Conference of the Parties to the NPT,' 24 May 2000, pp. 13–14.

[76] Pelzer, Dr. Norbert, 'The Nature and Scope of International Cooperation in Connection with the Peaceful Uses of Atomic Energy, and its Limits—an Assessment,' 1981, p. 138.

[77] Shaker, Mohamed, 'The Nuclear Non-Proliferation Treaty: Origin and Implementation 1959–1979,' 1980, p. 340.

The 1975 NPT Review Conference also encouraged increased assistance to developing countries both bilaterally and through multilateral channels.[78]

The 2000 Review Conference also noted that special needs and priorities of the least developed countries party to the treaty should be taken into account in bilateral and multilateral nuclear technical assistance and cooperation programmes.[79]

2.2. *Multilateral Regional Cooperation*

Multilateral regional cooperation has increased as a result of the NPT. Article IV is drafted in a way to encourage such cooperation. In Europe, for example, there are three organisations specifically engaged in nuclear energy cooperation, namely the EURATOM, the OECD NEA and CERN. Moreover, in the wake of the October 1973 Middle East war and the oil embargo, the International Energy Agency (IEA) was established in November 1974, as an autonomous body within the framework of the OECD, whose overall responsibility was to implement an International Energy Programme. The major objective of this programme was to reduce excessive dependence on oil and to develop alternative energy sources, including nuclear energy.[80]

On the other hand, African Countries under the auspices of the IAEA concluded an Agreement in 1990 for African regional cooperation (AFRA). The preamble of the Agreement stressed on the importance of mutual cooperation between States in the same region in the field of nuclear science and technology, and at the same time, recognised the importance of IAEA role in promoting cooperation among its Member States to assist their national atomic energy programmes.[81]

The 2000 Review Conference also recognised the importance of regional co-operative arrangements for the promotion of peaceful use of nuclear energy. It also noted the contributions of AFRA, the Regional Co-operative Agreement for the Advancement of Nuclear Science and Tech-

---

[78] Pelzer, Dr. Norbert, 'The Nature and Scope of International Cooperation in Connection with the Peaceful Uses of Atomic Energy, and its Limits—an Assessment,' 1981.

[79] NPT Review Conference, NPT/CONF.2000/28, 'Final Document of the 2000 Review Conference of the Parties to the NPT,' 24 May 2000, p. 14.

[80] Shaker, Mohamed, 'The Nuclear Non-Proliferation Treaty: Origin and Implementation 1959–1979,' 1980, pp. 342–343.

[81] IAEA, INFCIRC/377 'African Regional Co-operative Agreement for Research, Development and Training Related to Nuclear Science and Technology,' 2 April 1990.

nology for Asia and the Pacific (RCA), as well as the regional Technical Cooperation Programme in Central and Eastern Europe.[82]

Now, here it is important to distinguish between the different roles of some of those organisations and to see to what extent they complement each other.

EURATOM[83]

The EURATOM Treaty was signed in Rome in 1957. The main aim of the treaty is to undertake various measures that together promote and, to a lesser extent, to regulate nuclear power across Europe. The treaty is the legal basis for the EURATOM loans system, which provides subsidised credit for developing nuclear projects.[84] The EURATOM Treaty has two fundamental objectives: to ensure the establishment of basic installations necessary for the development of nuclear energy in the Community, and to ensure that all users in the Community receive a regular and equitable supply of ores and nuclear fuels. According to its Statute, EURATOM is entrusted with the task of contributing to raising the standard of living in the member States and to the development of relations with other countries by creating conditions necessary for the speedy establishment and growth of nuclear industries. Article 2 of the Statute provides that EURATOM shall:[85]

- Ensure that all users in the Community receive a regular and equitable supply of ores and nuclear fuels;
- Make certain, by appropriate supervision, that nuclear materials are not diverted to purposes other than those for which they are intended;
- Exercise the right of ownership conferred upon it with respect to special fissile materials;
- Ensure wide commercial outlets and access to the market in specialised materials and equipment, by free movement of capital investment in the field of nuclear energy and by freedom of employment for specialists within the Community.

---

82 NPT Review Conference, NPT/CONF.2000/28, 'Final Document of the 2000 Review Conference of the Parties to the NPT,' 24 May 2000, p. 15.

83 EU, European Union homepage.

84 Friends of the Earth Europe, 'Abolition of the EURATOM Treaty.'

85 Nuclear Energy Agency, 'The Regulation of Nuclear Trade: Non-Proliferation—Supply-Safety: International Aspects' Volume 2, 1988, pp. 29–30.

Article 52 of the EURATOM Treaty entrusted the commercial tasks relating to the conclusion of contracts among producers and users of ores, source materials or special fissile material to a separate body from the Commission itself, though placed under its control: the Supply Agency that is operating since 1960.[86] The mission of the Agency is to ensure a regular and equitable supply of nuclear fuels for Community users. It has legal personality and financial autonomy.[87]

The EURATOM Treaty provides the Supply Agency with the right of option to acquire ores, source materials and special fissile materials produced in the Community and an exclusive right to conclude contracts for supply of such materials from inside the Community or from outside. In order to be valid under Community law, supply contracts must be submitted to the Supply Agency for conclusion. The Supply Agency and the Commission pursue the objective of long term security of supply through a reasonable diversification of supply sources and the avoidance of excessive dependency on any one supply source, and ensure that in a context of fair trade, the viability of the nuclear fuel cycle industry is maintained.[88]

EURATOM not only co-ordinates national nuclear research programmes but also has its own research programme. Its Joint Nuclear Research Centre consists of four research centres located at Ispra (Italy), Petten (the Netherlands), Geel (Belgium) and Karlsruhe (FRG). Some research is also conducted by virtue of contracts signed with member States, enterprises or international organisations.[89]

NEA

The OECD NEA was established in 1957. The primary objective of NEA is to promote cooperation among governments of participating countries in order to further the development of nuclear power as a safe, environmentally acceptable and economic source. This is achieved by:[90]

---

[86] Ibid, p. 686.
[87] EU, European Union homepage.
[88] Ibid.
[89] Shaker, Mohamed, 'The Nuclear Non-Proliferation Treaty: Origin and Implementation 1959–1979,' 1980, p. 345.
[90] Nuclear Energy Agency, 'The Regulation of Nuclear Trade: Non-Proliferation—Supply-Safety: International Aspects,' Volume 2, 1988, p. 2.

- Encouraging harmonisation of national, regulatory policies and practices (with particular reference to the safety of nuclear installations) protecting humans against ionising radiation and preserving of the environment, managing radioactive waste and nuclear third party liability and insurance;
- Assessing the contribution of nuclear power to overall energy supply by keeping under demand and supply for different phases of the nuclear fuel cycle;
- Developing exchanges of scientific and technical information particularly through participation in common services;
- Setting up international research and development programmes and joint undertakings.

NEA supports efforts by member countries to improve their cooperation in the nuclear field by:[91]

- Supporting the administration, implementation and interpretation of the Paris and Brussels Supplementary Conventions and the modernisation of the international nuclear liability regime, primarily through the work of the Nuclear Law Committee;
- Providing legal services to operational activities and projects of the NEA;
- Analysing and disseminating, on a world-wide basis, information on nuclear legislation, jurisprudence, international agreements, and other relevant developments in the field of nuclear law;
- Assisting the countries of Central and Eastern Europe and the New Independent States in developing nuclear legislation that will facilitate their integration into the international nuclear liability regime;

CERN

This organisation was established in 1954. The founding States were Belgium, Denmark, France, Germany, Greece, Italy, the Netherlands, Norway, Sweden, Switzerland, the UK and Yugoslavia. Subsequently, Austria and Spain joined the Organisation. Both Yugoslavia and Spain withdrew later for financial reasons, but the former joined once more as an observer State (following the example of Turkey). The Headquarters of CERN are in Geneva.[92]

---

91 OECD homepage.

92 Shaker, Mohamed, 'The Nuclear Non-Proliferation Treaty: Origin and Implementation 1959–1979,' 1980, p. 348.

CERN is the largest particle physics centre in the world. The laboratory provides physicists with the tools necessary for exploring what matter is made of and what forces hold it together; among them accelerators, which accelerate particles almost to the speed of light, and detectors that make them visible.[93]

Article II of the 1953 Convention for the establishment of CERN stipulates that the organisation "shall provide for collaboration among European States in nuclear research of a pure scientific and fundamental character, and in research essentially related thereto. The organisation shall have no concern with work for military requirements and the results of its experimental and theoretical work shall be published ..."[94]

The main features of CERN are that it is a non-political organisation, its activities are of little concern to powerful industrial or commercial interests and that the national activities of the State members do not duplicate or compete with its work.[95]

AFRA

AFRA stems from an initiative of African member states who requested the IAEA in 1988 to assist them in establishing a regional arrangement for cooperation in the field of nuclear science and technology in Africa similar to the existing ones in Asia and Latin America. The Agreement entered into force on 4 April 1990 and today enjoys a membership of 29 African countries.[96]

AFRA is an intergovernmental agreement to promote the development and application of nuclear science and technology in Africa. It seeks to accelerate development towards self-sufficiency in nuclear scientific disciplines and relevant technologies in Africa. Co-ordinating intellectual and physical resources and implementing innovative methods and practices allows this to take please in a cost-effective manner.[97]

In 2000, AFRA member States approved the AFRA Vision to guide its cooperative activities during the third five-year term (2000–2005). IAEA has

---

[93] CERN homepage.

[94] 'Convention for the Establishment of a European Organisation for Nuclear Research,' 1 July 1953, amended on 17 January 1971.

[95] Shaker, Mohamed, 'The Nuclear Non-Proliferation Treaty: Origin and Implementation 1959–1979,' 1980, p. 348.

[96] AFRA homepage.

[97] Ibid.

the mandate to provide AFRA with technical and scientific backstopping, as well as financial and administrative support, in accordance with its rules and the procedures that govern the provision of technical assistance to its member states.[98]

OPEN

On 14 January 1974, a number of Belgian, French, German, Spanish and Swiss electricity producers formed OPEN (Organisation des Producteurs d'Energies Nucléaire) with legal status of a "Groupement d'Interet Economique" (GIE) whose membership is restricted to European electricity producers. The admission of new members is subject to unanimous consent of existing ones. The objectives of OPEN include:[99]

- Studying ways of satisfying the needs of their members with respect to separation of uranium isotopes, using, in particular, enrichment plants located in Europe;
- Undertaking measures designed to achieve optimum safety and economy for nuclear fuel supplies;
- Seeking and promoting all measures capable of improving, through mutual collaboration, procedures for managing nuclear fuels.[100]

To this end, the GIE is empowered:

- to study the detailed provisions of separation contracts and all provisions likely to stabilise the market or to facilitate financing separation plants;
- to establish procedures for supplying electricity to a separation plant using the gaseous diffusion process, by using an enrichment exchange in addition to electricity supply, for example, and draw up the reciprocal commitments of those members participating in this supply;
- to study detailed rules for sharing nuclear fuels and exchanging energy, so as to make supply contracts more flexible and reduce technical risks involved in constructing and operating power plants.

---

[98] Ibid.

[99] Shaker, Mohamed, 'The Nuclear Non-Proliferation Treaty: Origin and Implementation 1959–1979,' 1980, pp. 348–349.

[100] Nuclear Energy Agency, 'The Regulation of Nuclear Trade: Non-Proliferation—Supply-Safety: International Aspects,' Volume 2, 1988, pp. 34–35.

OPEN has also drafted an agreement, opened for signature by its members in September 1982, defining conditions upon which each signatory undertakes to make certain quantities of natural uranium available to the other signatories, as well as laying down the rules for loans and repayments resulting from these commitments.[101]

2.3. *International Cooperation through International Organisations*

At the Conference of NNWS, preference for international cooperation through international organisations was sometimes expressed either generally, without specifying a particular organisation, or without deciding between the use of existing organisations and the establishment of new machinery. Most of the participating States in the Conference opted for the IAEA, hence, they stressed the need for strengthening its role and reorganising it, in order to meet the expected intensification of international cooperation in peaceful uses of nuclear energy as a result of the NPT.[102]

IAEA

The Agency plays a major role in transferring nuclear technology for peaceful purposes through its Technical Cooperation Programme. The programme is developed jointly by the Secretariat and member States. It is based on the assessment of development priorities and conditions in each specific country or region, the project requests received from Member States, the application of appropriate criteria for project formulation, and finally appraisal and formal approval by the BG. The programme also includes regional and interregional projects that are developed to improve the efficiency of implementation or to better utilise the collective experience and resources of multiple member States.[103]

Due to the scarce resources available for technology transfer, the Programme focuses on activities that both contribute to national development and are cost-effective. This has led to a series of initiatives which, taken together, represent a gradual shift in emphasis in the Agency's Technical Cooperation Programme from project activities directed at

---

[101] Shaker, Mohamed, 'The Nuclear Non-Proliferation Treaty: Origin and Implementation 1959–1979,' 1980, pp. 348–349.

[102] Ibid, pp. 349–350.

[103] IAEA, 'The Technical Cooperation Programme.'

building capacity in nuclear authorities and institutions, towards collaboration with counterpart organisations to employ this capacity for productive and sustainable human development.[104]

In 1997, Technical Cooperation Strategy stated the new strategic goal for technical cooperation as follows: "Technical Cooperation with the Member States shall increasingly promote tangible socio-economic impact by contributing directly in a cost-effective manner to the achievement of the major sustainable development priorities of each country." The Agency specified that this cooperation take the form of partnership between it and each Member State, with the goal of achieving sustainable development. Technical Cooperation projects may comprise one or more of the following components: experts, equipment and materials, fellowships and scientific visits, training courses, meetings/workshops and sub-contracts.[105]

United Nations Development Programme (UNDP)

It is to be noted that apart from the IAEA, the UNDP, which is the main source of multilateral assistance for pilot plant and pre-investment project's playing a role in the supply of nuclear material and technology around the world.[106]

The administrator of the UNDP informed the UN SG that the programme funds can be used for pre-investment studies in: large scale nuclear energy, training, research, industrial advisory functions in the field of isotopes and ionising radiations, and technical and economic studies in the field of peaceful nuclear explosives, though only if this field reaches the stage of practical application in the case of the NNWS. The governing council, while considering requests for assistance in the field of atomic energy, would continue to be guided by its usual criteria of project soundness and priority. This includes the ability of the requesting government to provide the counterpart with appropriate facilities and qualified national staff, and requires the consideration of the likelihood of investment follow-up in appropriate cases.[107]

---

[104] IAEA, Board of Governors, GOV/INF/824, 'Technical Cooperation Strategy,' 1997, p. 1.

[105] Pelzer, Dr. Norbert, 'The Nature and Scope of International Cooperation in Connection with the Peaceful Uses of Atomic Energy, and Its Limits—An Assessment,' 1981.

[106] Shaker, Mohamed, 'The Nuclear Non-Proliferation Treaty: Origin and Implementation 1959–1979,' 1980, p. 355.

[107] Ibid, p. 370.

The DG of the IAEA at the time, Dr. Eklund, acknowledged that the large-scale projects the IAEA is executing on behalf of the UNDP "reflects the growing interest and capability of developing countries in carrying out larger projects in the nuclear energy field. It also indicates that the need for international assistance in this field is growing." In this connection, Dr. Elkund went on to say that "(t)he industrial countries, and particularly NWS, will certainly be called upon to fulfil the commitments they have made in Article IV of the Treaty."[108]

To conclude this brief review of the role of the UNDP in promoting the peaceful uses of nuclear energy, it must be said that developing countries must take full advantage of the funds available for technical assistance within the UNDP programme. As noted by IAEA, nuclear energy projects are in many cases relegated to such a low priority that no such projects can be carried out with the UNDP resources available for those countries. As to the large-scale projects, it is suggested that developing countries should review their requirements to ascertain whether they wish to give higher priority to nuclear projects, especially of the kind already successfully executed in other countries. On the other hand, advanced countries should make more funds available so that the UNDP may be able to cope with future requests for technical assistance and large-scale projects.[109]

3. *US Regulations and International Cooperation*

Transfer of nuclear technology to beneficiary states depends to a great extent on national regulations of Supplier States. Since the establishment of the IAEA and the NPT, Supplier States were encouraged to keep on transferring nuclear technology for peaceful purposes to NNWS for the benefit of humankind. It was important for national regulations of Supplier States to comply with the international rules for non-proliferation, especially if the Beneficiary State is not a member of the NPT and its nuclear facilities are not under IAEA safeguards.

The importance of national regulations was recognised in the Review Conferences of the NPT. The 2000 Review Conference recognised that national rules and regulations of State parties are necessary to ensure that they are able to fulfil their commitments with respect to the transfer of nuclear and nuclear-related dual-use items to other States.

---

108 Ibid, p. 371.
109 Ibid.

It also encouraged nuclear States to follow the spirit of the NPT with respect to technical cooperation with States not party to the treaty.

3.1. US *National Rules*

It seems reasonably certain that large-scale development of peaceful atomic energy in economically underdeveloped areas is not an imminent prospect, and that the measured pace of the American power reactor programme will not cause the "Atoms for Peace" programme to fail. It was foreseen that IAEA, and regional centres, were indubitably worthwhile institutions. But, for decades, their usefulness would be largely confined to giving technical assistance, and especially to helping with training, information, and project analysis.[110] This statement is true to a great extent. The IAEA role today more than 45 years after this comment, while promoting the use of nuclear energy for peaceful purposes, is confined to the previously mentioned technical cooperation programmes.

At first glance, IAEA certainly looks like a better channel of supply to underdeveloped countries than bilateral programmes. However, a decision to substitute the IAEA completely for the bilateral programme is not, of course, without drawbacks. American aid is greater under bilateral than multilateral arrangements. Yet, the IAEA would seem the superior vehicle, provided the aid involved is not a great deal more than technical assistance.[111]

Nevertheless, the US is still an appealing supplier of technology and materials. By mid-1974, the Congress realised that US foreign and energy policies were not sufficient to cope with nuclear weapons proliferation. Accordingly, after the 11 October 1974, enactment of the Energy Reorganisation Act of 1974, which abolished the Atomic Energy Commission (AEC) and created the Energy Research and Development Administration (ERDA) and the Nuclear Regulatory Commission (NRC), Congress amended section 123 of the Atomic Energy Act. The Act provided a modified procedure for closer Congressional scrutiny of bilateral agreements for the sale of nuclear reactors abroad.[112]

---

[110] The American Assembly, 'Atoms for Power: US policy in Atomic Energy Development,' December 1957, pp. 121–122.

[111] Ibid, p. 123.

[112] Doub, William O. and Eugene R. Fidell, 'International Relations and Nuclear Commerce: Developments in United States Policy,' Law and Policy in International Business, volume 8, number 4, 1976, p. 915.

India's successful nuclear test in 1974 shattered the US Government's laissez-faire nuclear export policy. The 1973–1974 oil embargo contributed to widespread expectations of an expanded role for peaceful uses of nuclear power. Supplier nations signed contracts with developing countries for facilities that could have provided those countries with fissile material.[113]

After the Indian nuclear tests, Doub & Fidell, mentioned that "In retrospect, it would not be irrational to contend that we [*Americans*] have given our nuclear technology assets away and received nothing in return. At the same time, no one questions that the Eisenhower "Atoms for Peace" programme was a success by its own standards: A world war produced the momentum for the development of civilian nuclear power and eventually inspired a dedicated commitment to make the benefits of this technology available on a world-wide basis. However, the events of 1974, although reflecting the increased potential for enjoyment of the peaceful uses of nuclear technology, also initiated the demise of the policies behind the "Atoms for Peace" programme. India's explosion of a nuclear device on 17 May 1974, came as a reminder that nuclear weapons capability may be well within the grasp of Third World powers. That the nuclear device had been fabricated from material produced in a Canadian supplied reactor not subject to international safeguards was perhaps small comfort to Congress and the American public. It is not unfair to say that the Indian explosion set back nuclear trade. It also brought world recognition of nuclear trade policies, and helped to focus public and governmental attention on the problems of nuclear proliferation under international and bilateral agreements."[114]

American Efforts to Create a Nuclear Export Policy

Alarmed at the events of the oil embargo of 1973–1974 and at the apparent acquiescence of the Ford Administration to the French-Pakistani deal, where Pakistan was to buy reprocessing facilities from France, the Congress intervened and passed the Symington Amendment.

[113] Cassidy, Stephen H., 'The Newest Member of the Nuclear Club: Pakistan's Drive for a Nuclear Weapons Capability and US Nuclear Non-proliferation Policy,' Hastings International Comparative Law Review, Volume 12, no. 1,1988/1989, pp. 688–689.

[114] Doub, William O. and Eugene R. Fidell, 'International Relations and Nuclear Commerce: Developments in United States Policy,' 1976, p. 916.

Symington Amendment

The Symington Amendment added section 669 to the Foreign Assistance Act of 1961. It mandated a cut-off of the US military and economic assistance to any nation that exported or imported nuclear reprocessing or enrichment facilities, unless that nation accepted full-scope IAEA safeguards.[115]

As 1974 drew to a close, despite two decades of reliability on supply, which characterised American foreign nuclear trade, the uncertainty and dislocation in US nuclear export policies were evident throughout the world. As a result, the US was unable to negotiate any new bilateral agreements for cooperation. The futility of the efforts exerted by the Government to continue its foreign nuclear programme is epitomised by its lengthy efforts to negotiate a mutually satisfactory agreement for cooperation with Iran. The Department of State and the Energy Research and Development Administration pursued comprehensive negotiations with Iran, and in 1974 a draft agreement was submitted to Iran for its consideration. As a result of the friction between Iran and its Arab neighbours over the 1973 Arab oil embargo; Iran met every prior test to qualify to pursue a commercial nuclear reactor programme. For years, it had been a staunch ally of the US, a stabilising force in the Middle East, a major (if discreet) supplier of petroleum to Israel, and a party to the NPT. In an effort to anticipate the closer Congressional review of new bilateral agreements, the Administration prepared a draft which differed remarkably from those previously executed (such as the Agreement for Cooperation between the US and Spain, made effective on 28 June 1974). However, by the middle of 1975, Iran decided that its order for two reactors should be given to West Germany, and then began negotiations with France for additional reactors. Iran was not the only buyer to shop elsewhere other than the US. In June 1974, West Germany signed a comprehensive agreement to supply Brazil with a complete nuclear fuel cycle, including reprocessing and enrichment. South Korea began negotiations for the purchase of a reprocessing plant from France, which also offered to sell a reprocessing plant to Pakistan after the US turned down Pakistan's request.[116]

---

[115] Cassidy, Stephen, 'The Newest Member of the Nuclear Club: Pakistan's Drive for a Nuclear Weapons Capability and US Nuclear Non-Proliferation Policy,' 1988/1989, p. 694.

[116] Doub, William O. and Eugene R. Fidell, 'International Relations and Nuclear Commerce: Developments in United States Policy,' 1976, pp. 917–919.

With the advent of 1976 and the continued instability of US export policy, Secretary of State Kissinger launched an aggressive campaign to void the sales of reprocessing plants to South Korea and Pakistan, and diplomatic pressure was used to "improve" the comprehensive sales package between West Germany and Brazil. The US reaction to these developments, however, was not limited to diplomatic pressure. In fact, the major component was a legislative effort to create a nuclear export policy designed to restore the US to a position as a major supplier, while simultaneously requiring close scrutiny of the impact of those exports upon foreign policy and military security.[117]

American Agencies Involved in Nuclear Trade. Trade in nuclear materials and technology within the US involves several agencies. The NRC licenses exports of nuclear facilities, materials and components, while the Department of Energy (DOE) controls technology exports and bilateral and multilateral technical exchange agreements. The DOE also assumes primary responsibility for "subsequent arrangements" pursuant to cooperation agreements; for example, approval of reprocessing of US-supplied nuclear materials or re-transfer of US-supplied materials among foreign nations. The Department of Commerce licenses exports of dual-use equipment and the balance of plant items related to nuclear facilities. Procedures for executive branch judgements on exports involve the Departments of State, Defence, Energy, and Commerce and the Arms Control and Disarmament Agency, as well as the NRC under the Nuclear Non-Proliferation Act (NNPA) of 1978.[118]

The NRC licenses exports of reactors, isotope separation plants, reprocessing plants, heavy water production facilities, plants for fabrication of nuclear reactor fuel elements, various components for the above facilities, special nuclear material, source material, by-product material, deuterium and nuclear grade graphite. The DOE also implemented special rules for technical knowledge, design, fabrication, and operation of such facilities. Technical data includes information of any kind such as a model, prototype, blueprint, or operating manual, as well as intangible technical services. To decide whether to grant a license, the DOE and the Department of Commerce consider the stated end use and significance of commodity or technical data, availability from non-US sources, guarantees against

[117] Ibid, p. 919.

[118] Nuclear Energy Agency, 'Nuclear Legislation, Analytical Study, Regulation and Framework for Nuclear Activities,' Volume II, 1984, pp. 177–178.

use for explosive purposes or proliferation, the participation of a country in international energy agencies and intelligence data on its nuclear intentions and activities. The exporter, responsible for licence application, must ask the manufacturer for relevant data when deciding if such an application is necessary.[119]

The NNPA requires the President to initiate discussions with supplier and recipient nations in order to develop new international approaches to nuclear energy. By co-ordinating with the IAEA, the DOE initiates training for foreigners to disseminate new safeguards techniques. The Foreign Assistance Act of 1961, as amended, precludes economic assistance to countries which deliver nuclear enrichment equipment or technology to other countries, unless the recipient country co-operates with multilateral and international safeguards agreements. A recipient country subject to the Act cannot obtain such materials from third countries without placing them under multinational control subject to IAEA safeguards. The President *of the US*, nevertheless, can furnish assistance otherwise prohibited if termination would be seriously prejudicial to the achievement of the non-proliferation objectives of the Government, and provided that the recipient State transmits reliable assurances that it will not acquire or develop nuclear weapons.[120]

Glenn Amendment

In 1975, the Glenn Amendment revised the Symington Amendment. The Glenn Amendment created section 670 in the Act to separate the provisions relating to enrichment from the language on reprocessing, and to include new language on the detonation of nuclear weapons. The Glenn-Symington Amendments, in turn, were amended by the International Security and Development Cooperation Act of 1981 to add Congressional vetoes by concurrent resolution and to provide a separate and strengthened subsection on the testing of nuclear devices.[121]

Section 669 governs transfer of nuclear enrichment technologies and comprises two subsections. The first subsection states that no funds, under the Foreign Assistance Act, may be provided to any nation that on or after 4 August 1977 delivers nuclear enrichment equipment, materials,

119 Ibid.

120 Ibid.

121 Cassidy, Stephen, 'The Newest Member of the Nuclear Club: Pakistan's Drive for a Nuclear Weapons Capability and US Nuclear Non-Proliferation Policy,' 1988/1989, pp. 694–695.

or technology from another country, unless before delivery the supplier and recipient countries place these items under multilateral auspices and the recipient country enters into a full scope safeguards agreement with the IAEA. The second subsection includes two key provisions. First, the President can waive the cut-off if he certifies in writing to the Congress that the termination of such assistance will have a "serious adverse effect on vital US interest[s]" and "he has received reliable assurances that the country in question will not acquire or develop nuclear weapons." Second, if within thirty calendar days after receiving certification, the Congress disapproves furnishing assistance by a concurrent resolution, the certification will cease to be effective and all aid will end.[122]

Section 670 of the Foreign Assistance Act regulates plutonium reprocessing transfers and receipt or use of nuclear weapons by foreign nations. Subsection (a) stipulates that no funds under the Act may be provided to any country, which on or after 4 August 1977 delivers or receives nuclear reprocessing equipment, materials, or technology from any other country, except as part of an international examination of alternatives to pure plutonium reprocessing in which the US participates. The President may continue aid if he certifies that "termination of such assistance would be seriously prejudicial to the achievement of US non-proliferation objectives or otherwise jeopardise the common defence and security." The Congress can nullify the certification, however, by a concurrent resolution disapproving of a resumption of assistance. Subsection (b) of section 670 states that no US assistance may be provided to any nation which "(A) transfers a nuclear explosive device to an NNWS, or (B) is an NNWS and either—(i) receives a nuclear explosive device, or (ii) detonates a nuclear explosive device." The President may, for not more than thirty days of continuous session, furnish assistance that would otherwise be cut off if he transmits to Congress "a certification that he has determined that an immediate termination of assistance to that country would be detrimental to the national security of the US." Aid could not continue past that thirty-day period unless approved by a joint resolution of Congress and upon the same Presidential certification as contained in section 670 (a).[123]

A comparison between the cut-off provisions of sections 699 and 670 (a) reveals that the language of section 699 relating to nuclear enrichment

[122] Ibid.
[123] Ibid.

transfers is less restrictive than the language of section 670 (a) concerning reprocessing transfers. Nuclear enrichment transfers will not trigger termination of US assistance to nations engaged in the exchange if the recipient nation agrees to place enrichment components, along with all its nuclear fuel and facilities, under the IAEA safeguards. Section 670 (a) does not contain a similar exemption. The only exception to a cut-off for receipt or delivery of reprocessing technology is if the transfer is part of an international effort, in which the US participates, to find alternatives to plutonium reprocessing. The stricter language on plutonium reprocessing reflects the hypothesis held in Congress and by the Ford and Carter Administrations that plutonium reprocessing posed significant dangers to the non-proliferation regime. As a practical matter, the broader exception of Section 669 will probably never be applied. A nation seeking weapons capability and importing enrichment technology would not agree to full scope IAEA safeguards in order to avoid an aid cut-off.

The nations posing the greatest proliferation risks today have rejected the NPT and operate nuclear facilities that are not subject to IAEA safeguards. Section 669 and 670 (a) grant Congress the authority to revoke a Presidential certification by concurrent resolution. A concurrent resolution is a form of legislative veto. It requires only that both chambers of Congress approve the legislation before it becomes law. By virtue of the Supreme Court's holding in *Immigration and Naturalisation Service v. Chadha*, the concurrent resolutions are undoubtedly unconstitutional. Even so, the Congress could refuse to appropriate aid in subsequent fiscal years. *Chadha* does not affect section 670 (b), relating to receipt or use of a nuclear device, because the section does not contain a legislative veto. The section itself limits the authority of the President to override an aid cut-off.

The Congress determined that a transfer of an atomic device or explosion by an NNWS would be so grave an event that the President can only delay the cut-off of assistance for thirty days. After thirty days, the aid would automatically expire. It could only be restored if the President certified to Congress that termination of assistance would be seriously prejudicial to US interests and both Houses of Congress approved the assistance. Hence, after the endorsement of the Symington Amendment, President Ford in October 1976 announced that the US would defer the reprocessing and recycling of plutonium from civilian nuclear power plants. The decision generated shock waves throughout nuclear programmes world-wide. If commercial reprocessing was not economic in the world's largest nuclear power producing state, it was absurd for Pakistan. European and Japanese policymakers feared that the US was seeking to sabotage their plans for plutonium reprocessing and development in breeder reactors.[124]

Ironically, the experience with Pakistan has proved the Presidential Certification Provision of section 669 to be more rigorous than the provision contained in section 670 (a). To override an aid cut-off triggered by a nuclear enrichment transfer, the President must certify to Congress both that the termination of aid would have seriously adverse effects on vital US interests and that he has received "reliable assurances" that the recipient nation is not developing nuclear weapons. To continue US assistance after a nuclear reprocessing transfer, the President needs only to certify that US interests would be seriously prejudiced by an aid cut-off. Because the Reagan Administration could not receive "reliable assurances" from the Government of President Zia-ul Haq that Pakistan was not developing nuclear weapons, the Administration was forced to request from Congress specific waivers of section 699 in 1981 and 1987 as part of its two aid packages for Pakistan.[125] The waivers were granted and the flow of US nuclear assistance to Pakistan continued.

Carter's Policy

In April 1977, President Carter announced his Administration's policies regarding proliferation and nuclear power. The US would now indefinitely defer commercial reprocessing and recycling of plutonium and continue to embargo the export of enrichment and reprocessing technology. President Carter also called for the establishment of an international fuel cycle evaluation to deal with nuclear problems adding a sense of urgency to non-proliferation issues. Title I states that it shall be US national policy to take such action as necessary for the US to provide a reliable supply of nuclear fuel. Title II commits the US to strengthen IAEA safeguards. Title III amends the AEA of 1954 by requiring an export licence, based upon certain statutory criteria, for US nuclear assistance and supplies. Among the criteria for an export licence, section 305 of Title III provides that the recipient nation pledges not to use US material for the development of a nuclear explosive device, not to reprocess or transfer US supplied fuel without prior approval, and to place all of their peaceful nuclear activities

---

[124] Ibid, pp. 697–698.
[125] Ibid.

under IAEA safeguards. Furthermore, the Act imposed upon the President the obligation to renegotiate all existing nuclear cooperation agreements to be consistent with the export criteria of the Act.[126]

Concern over nuclear weapon proliferation prompted the President of the US to defer indefinitely commercial reprocessing and recycling of plutonium produced in US nuclear power programmes. This paved the way for the US NNPA of 10 March 1978. The Act was a compromise between provisions meant to strengthen US reliability as a supplier of commercial nuclear technology, and requirements to impose severe unilateral conditions on American nuclear exports.[127] It prohibits the export of sensitive nuclear material and equipment to NNWS, such as plutonium and reprocessing plants, and requires tightening of international safeguards on recipient countries. The US has even decided for some time not to export sensitive facilities and technologies. The new or renegotiated cooperation agreements would stipulate that the latter would not be transferred unless the agreements are specifically amended to provide for them. Paradoxically, this obvious over-concern with the proliferation of nuclear weapons runs counter to the letter and spirit of Article IV of the NPT and more specifically with the pledges undertaken in its second paragraph.[128]

The US policies addressed regional and bilateral nuclear supply arrangements. However, it did not grant priority to NNWS parties to the NPT. It made arrangements with States, not NPT parties, as long as the supplied project was in conformity with its interests (be the project political, economic or for research purposes), regardless of the NPT position, thereof giving priority to NNWS parties to the NPT.[129] Hence, a NWS disregarded the NPT rules while engaging in international nuclear cooperation agreements. Whether this is a consequence of the insufficiency of the rules

[126] Ibid, p. 699.

[127] Ibid.

[128] Shaker, Mohamed, 'The Nuclear Non-Proliferation Treaty: Origin and Implementation 1959–1979,' 1980, p. 290.

[129] Ibid, p. 410. In 1969, Senator Fulbright, the then Chairman of the Senate Committee on Foreign Relations, criticised the "Project Kerauderen" for Australia, whether it was not "a little odd" to choose a country which has not signed the NPT (it signed it only on 27 February 1970) while there were other countries which could be favoured. The clarification given by the then Chairman of the US Atomic Energy Commission, Dr. Seaborg, was that the Australian project was not an actual application of peaceful nuclear explosions but a research and development project which conformed to the AEC's own programmatic interests and replaced an experiment which it would otherwise have done elsewhere.

to satisfy the interests of the NWS, or its political interests supersede the rules, the outcome is alike. Both represent uncertainties in implementing the rules of the treaty.

Reagan's Policy

President Reagan established his Administration's non-proliferation policy for his entire term of office in a speech delivered on 16 July 1981. The speech delineated "basic guidelines" for US policy:[130]

- Prevent the spread of nuclear weapons as a fundamental national security and foreign policy objective;
- Strive to reduce the motivation for acquiring nuclear weapons and to promote regional and global stability and the legitimate security needs of other states;
- Continue to support adherence to the NPT and work to strengthen the IAEA safeguards system;
- Continue to inhibit the transfer of sensitive nuclear material, equipment and technology and to seek full-scope IAEA safeguards as a condition on new nuclear supply commitments;
- Re-establish the US as a reliable nuclear trading partner under adequate safeguards; and
- Co-operate with other nations by not inhibiting foreign commercial reprocessing and breeder reactor development in nations with advanced nuclear power programmes where it does not constitute a proliferation risk.

In October 1981, President Reagan released a statement on domestic nuclear power with implications for non-proliferation policy. President Reagan lifted the indefinite ban on commercial reprocessing activities in the US and strongly endorsed commercial use of plutonium in nuclear power generation and the development of breeder reactor technology. Due to Soviet invasion of Afghanistan, the Reagan Administration set an aid package to Pakistan. The stated purpose of the programme was "to benefit the people of Pakistan by helping them meet the burdens imposed by the presence of Soviet forces in Afghanistan and by promoting economic development."[131]

[130] Cassidy, Stephen, 'The Newest Member of the Nuclear Club: Pakistan's Drive for a Nuclear Weapons Capability and US Nuclear Non-Proliferation Policy,' 1988/1989, p. 701.
[131] Ibid, p. 712.

On 10 February 1982, President Reagan exercised the authority granted to him by the International Security Act to waive the prohibition of section 669 of the Foreign Assistance Act for Pakistan, related to transferring nuclear enrichment technology from 29 December 1981 until 30 September 1987. President Reagan also waived provisions of section 670 (a) of the Foreign Assistance Act with respect to a cut-off of funds for receipt of plutonium reprocessing technology by utilising the "built-in" waiver of the section. Though principally designed to serve US strategic interests in the region, the Reagan administration also argued that US assistance would diminish the Pakistani motivations to build a bomb. However, Pakistan's nuclear intentions had been clear to Washington for years and by virtue of President Reagan's action on 12 February 1982, Pakistan could continue its efforts to obtain weapons capability without risking an American aid cut-off, provided it did not test a nuclear device. The US was left only with export controls and diplomatic pressure on Pakistan as primary tools to curb the Pakistani nuclear programme.[132]

Solarz Amendment

During 1985, another amendment was added to section 670 (a) of the Foreign Assistance Act called the Solarz Amendment. Section 670 (a) included prohibitions on transfers of nuclear reprocessing technology and mandated an aid cut-off if an NNWS, its agent, or a person otherwise acting on behalf of the nation, on or after 8 August 1985 exports illegally (or attempts to export illegally) from the US any material, equipment, or technology which would contribute significantly to the ability of such country to manufacture a nuclear explosive device, if the President determines that material, equipment, or technology were to be used by such country in the manufacture of a nuclear explosive device. This amendment is less restrictive than the other cut-off provisions of sections 669 and 670 (a) and (b) because it is not self-executing. The President must first determine that the nation in question was involved in the illegal effort. Under the other cut-off provisions, the authority of the President to furnish assistance expires upon the triggering event. As in case of plutonium transfers, the President may definitely waive the Solarz Amendment through certification to Congress that "termination of such

132 Ibid, p. 713.

assistance would be seriously prejudicial to the achievement of US non-proliferation objectives or otherwise jeopardise the common defence and security."[133]

President Reagan waived, by certification to Congress, three of the four cut-off provisions of the Foreign Assistance Act which apply to all nations receiving US assistance. On 15 January 1988, he waived until 1 April 1990, section 669, which denies aid to any nation receiving unsafeguarded nuclear enrichment technology. On 10 February 1982, President Reagan indefinitely waived section 670 (a) (1) (A), relating to transfers of nuclear reprocessing technology, pursuant to the "built-in" certification provision of section 670 (a)(2) of the Act. Also, pursuant to section 670 (a)(2), the President waived section 670 (a)(1)(B), the Solarz amendment applied only retroactively to activities of Arshad Perez on behalf of Pakistan. A waiver of the remaining cut-off provision, section 670 (b), has not been considered since it penalises nations involved in transfer or detonation of nuclear weapons by NNWS.[134] Pakistan had neither received nor tested a nuclear weapon at the time. It tested a nuclear explosive device at a much later date in the 90s.

Effect of Foreign Policy Priorities on US Nuclear Policy

US governmental policy considers nuclear proliferation, the spread of the capability to test and produce nuclear weapons in NNWS, to be a grave threat to global peace and stability. The acquisition of atomic bombs by any nation has enormous implications. The development of imbalances in military power and the creation of nuclear arms races are primary concerns. Other dangers include greater opportunities for terrorist attacks and accidental detonation, the rapid escalation of conventional armed conflict into nuclear confrontations, and the possibility that unstable and provocative national leaders would actually seek to use nuclear weapons.[135]

The USSR invasion of Afghanistan in 1979 resulted in a shift in US foreign policy priorities. Previous Carter administration goals of conventional arms restraint, discouragement of security commitments in Third

---

[133] Ibid, pp. 717–718.

[134] Cassidy, Stephen, 'The Newest Member of the Nuclear Club: Pakistan's Drive for a Nuclear Weapons Capability and US Nuclear Non-Proliferation Policy,' 1988/1989, pp. 729–730.

[135] Ibid, pp. 681–682.

World, and ostracism of nations engaged in nuclear weapons proliferation gave way to a re-affirmation of the US defence treaty with Pakistan by President Carter and an offer of 400 million dollars in economic and military aid. US policymakers hoped this economic and military assistance would, in part, curb the Pakistani desire for nuclear weapons capability. Within a few years, it was apparent that the Pakistani nuclear programme was continuing unabated. The US, however, decided to renew the aid packages in 1987. It was perceived to be in the US national interest to provide funds for Pakistan, despite non-proliferation concerns, in return for Pakistan's support of the Afghan rebels resisting the Soviet occupation of Afghanistan. Today, the nuclear rivalry in South Asia poses one of the greatest challenges for US nuclear non-proliferation policy. The ability of the US to influence another nation's nuclear policy—the concept of leverage—depends upon a number of factors including the following: level of technological capability, scientific infrastructure, governmental stability needs, economic well-being, and political relationships.[136]

However, since the late 90s when the world overcame the shock caused by the Indian and Pakistani nuclear explosions, it became an established fact that they are both declared NWS, but the US is mute on this issue. It focuses nowadays on Iran—which the accusations against it so far have not been proved in accordance with the IAEA reports which stated that no nuclear weapons or nuclear production programmes were uncovered in Iran, but the IAEA needs to clarify certain aspects of the Iranian programme to provide it with a clean. Nevertheless, North Korea, which withdrew from the NPT and is known to the world to have produced nuclear weapons, is facing milder political pressures than Iran. Hence, the fact remains that the ones who have the Bomb are safe and US non-proliferation policy stands crippled before such cases.

3.2. *Effects of US Nuclear Policy on Nuclear Proliferation*

In the Third World, US opposition can delay, but not indefinitely prevent, a nation, that has a moderately developed scientific infrastructure and the necessary political will, from developing a nuclear weapons capability. Where a nation has a limited scientific base, as in the case of Libya,

---

[136] Ibid, pp. 685–686.

cutting off technology can be effective. In some situations, however, it may result in the loss of influence and temporarily strengthen local "pro-bomb" advocates.[137]

As nuclear capabilities spread and American influence on international energy and security policies diminishes, the political dimension of nuclear proliferation assumes a pivotal role. Technical approaches such as export embargoes, and institutional solutions (principally IAEA safeguards), are fundamental components of the non-proliferation regime. However, these measures alone cannot halt proliferation because the decision to manufacture a nuclear weapon is essentially a political one.[138]

3.3. *Constraints on US Non-Proliferation Policy*

It is important to emphasise that specific constraints on any nation's non-proliferation policy exist. First, general approaches must be adapted to individual nations. It is unrealistic and potentially counterproductive, to apply a blanket non-proliferation policy uniformly to all nations. Nations have unique histories, security interests, nuclear motivations, capabilities, and political situations. For example, the construction of a reprocessing plant in Japan would not evidence a covert programme to obtain nuclear explosive- or fissile-material, whereas with Pakistan it would and it did. This is an argument used to hamper the right of NNWS to access nuclear technology. The second constraint is that proliferation includes many steps, namely, purchase of commercial nuclear power plants or constructing facilities for the production of fissile material, followed by the testing of an atomic bomb. The third caveat concerns the unpredictability of the near future. President Kennedy envisaged the possibility that there would be fifteen to twenty five nuclear weapons states by the 1970s. Chance is certainly one of the elements that contributed to the slow rate of proliferation and the non-use of nuclear weapons after 1945. But, the Iraqi use of chemical weapons in its war against Iran illustrates how decades of restraint and international convention can evaporate overnight. High interest rates, world recession, low electricity demand, falling oil prices, and skyrocketing construction costs all left the most respected predictions of the 1970s concerning the growth of nuclear industry worthless by the mid-1980s. The last caveat is the

---

[137] Ibid, pp. 686–687.
[138] Ibid, p. 687.

consideration of foreign policy objectives. In the mid-1970s, several circumstances combined to make non-proliferation one of the highest US foreign policy priorities.[139]

3.4. *US Concerns Related to Nuclear Exports*

The US had expressed great concern about West Germany's sale to of a complete fuel cycle to Brazil, but in West Germany, the nuclear power programme was snarled in controversies about waste management and reactor safety.

Although the IAEA provides international co-ordination on environmental issues, there is often an assumption in policy making that these environmental issues are purely local and that it is up to each country to protect its own environment. It is important to place some of these pressing environmental issues in a global context, to recognise that one country's decisions and actions may affect the decisions and actions of other countries, and to understand that there is a common interest in the safe handling of this dangerous technology. Schachter stated that "while examining the three examples of reactor safety, waste management and transportation of spent fuel, the following can be underlined:

- Reactor safety is most often seen as the kind of issue that is purely local. The official US position is to take a hands-off approach with regard to safety in the export process. The reactor safety problem has international implications. However credible or incredible accidents may be in the US, they may be more credible abroad, particularly in developing countries, which may not have the same maintenance and repair capability that we [the Americans] have here. The Indian situation at the Tarapur Atomic Power Station is a case in point. In 1973, an AEC official remarked that, based upon what he knows of the Tarapur Atomic Power Station, it was a prime candidate for nuclear disaster. The effects of a nuclear disaster in India or Taiwan or elsewhere are global in at least two senses. First, such an incident would have far-reaching ramifications for nuclear industry world-wide. Indeed, such an accident could well spell the end of nuclear industry. Second, the damage that would be brought by such an accident might well not be limited to a single country or group of countries.

[139] Ibid, p. 688.

– Waste management is another issue of global consequences ... Already, both former President Ford and President Carter have emphasised the possibility in exploring international spent fuel storage regimes in order to reduce proliferation risks. But such regimes also may be significant from an environmental point of view, for they may provide a mechanism for safe, long-term handling of waste which simply does not now and may never exist in many individual countries.

Transport of radioactive wastes is a third example of an area of concern with broad international implications. If we move to an international waste storage regime or an alternative regime involving transfer of spent fuel to NWS generally or return of spent fuel to the US, or real alternatives, and if nuclear industry continues to grow, there will be a vast number of shipments or radioactive wastes around the world."[140]

While some international and multinational regulations would be appropriate, a medley of national and multinational initiatives are required; also although it has often been stated that one country cannot impose its own health, safety, environmental standards on others, there is a real opportunity for bilateral cooperation in nuclear trade, and the exercises of what could be termed "supplier responsibility" with respect to health, safety, environmental matters. Contrary to the official position taken by the US Government, there is, in a number of cases, a real desire on the part of recipient countries to benefit from US expertise. There is also a real desire on the part of the public in countries that have co-operative trading relations with the US to obtain adequate information about the risks and benefits of nuclear power, information which they simply could not obtain from their own governments. It was the conclusion of the Ford Foundation study that the current lack of control over the safety of American nuclear exports is a major deficiency in the programme. In sum, if the environmental and safety risks are to be eliminated or reduced, there is a need not only to strengthen existing international institutions, but also to recognise that, at least in some interim period, and perhaps over a relatively long time, there will be a need for supplier countries like the US to take a firm stance on a variety of health, safety, and environmental

[140] Schachter, Oscar, 'United Nations Law,' American Journal of International Law, Volume 88, 1994, pp. 416–417.

issues. The Ford Foundation study stated that there is no US Agency with responsibility to assure safety standards for exported reactors.[141]

4. *Conclusion*

To conclude in assessing the US rules in transferring nuclear technology, the following can be highlighted:

1. US rules allow the transfer of nuclear technology for peaceful purposes. Though it has dominated the market for a long time, it has never been concerned by any factors like nuclear safety or waste management in beneficiary States. However, its main concern focused on how to ensure that beneficiary States would not divert the purpose of the transferred technology from peaceful to military.
2. After the Indian test of a nuclear explosive device in 1974, the US started reforming its rules to ensure that no technology or materials imported from the US could be used as part of a nuclear programme for military purposes in an NNWS.
3. In transferring nuclear technology, US rules gave priority to NNWS members of the NPT and whose facilities are under IAEA safeguards. However, practice proved otherwise. After the Soviet invasion of Afghanistan, Pakistan benefited by receiving nuclear assistance from the US for political reason. Pakistan was never a member of the NPT and its major nuclear facilities are not under IAEA safeguards. It was engaged in a programme to develop nuclear weapons when it was receiving nuclear assistance from the US. The American Administration was aware of this military programme; nevertheless, consecutive American Presidents waived the prohibition under NNPA and the flow of US nuclear assistance aid packages to Pakistan resumed. US interests (economic, political, or even for R&D purposes) are the main determinants of its nuclear policies internationally.
4. The US has exerted great efforts to halt the proliferation of nuclear weapons, internationally—by supporting the establishment of the IAEA and the conclusion of the NPT—and internally, through reforms it has undertaken in its national legal system governing the

[141] Ibid, pp. 419–421.

transfer of nuclear technology. However, political interests superseded the rules. The US not only disrupted all its global non-proliferation efforts, but also in some ways assisted an NNWS to become nuclear.

CHAPTER THREE

SUPPLY OF NUCLEAR MATERIAL AND EQUIPMENT

Ever since man mastered the energy released by atomic fission, a cloud has weighed on the peaceful uses of this source of energy—the possibility of their being diverted to the production of atomic weapons.[1]

1. *Introduction*

The movement of nuclear material and technology around the world constitutes an important factor that should be addressed, including its trade and the measures that the suppliers have taken so far, nationally and internationally, to prevent its misuse.

The entry into force of the NPT in 1970 has not managed to remove all concerns aroused by the proliferation risks associated with nuclear trade. Supplier States negotiated to specify conditions for applying safeguards. They adopted a policy of strengthening controls and placing embargoes on certain products and equipment. This policy disturbed importing countries, anxious to secure their supply sources and to avoid jeopardising legitimate desires to conduct their civil nuclear programmes successfully.[2]

The non-proliferation and the nuclear supplier regimes are defined by the policy of "haves" and "have-nots". At present, the policies of nuclear trade and cooperation pursued by the major nuclear suppliers reflect various interests, influences, informal and formal undertakings and commitments. Among these, the formal commitments of the NPT and the informal understandings and voluntary undertakings of the Zangger Committee and the London Supplier Group (NSG) guidelines are the common (and dominant) determinants of policy (if not also of behaviour).[3]

---

[1] Nuclear Energy Agency, 'The Regulation of Nuclear Trade: Non-Proliferation—Supply-Safety: International Aspects,' Volume 2, 1988, p. 9.

[2] Ibid, pp. 9–10.

[3] Pilat, Josef, 'The Major Suppliers: A Baseline for Comparison,' in William C. Potter (ed.), 'International Trade and Non-Proliferation: The Challenge of the Emerging Suppliers,' 1990, p. 39.

The informal Zangger Committee, named after its Swiss Chairman, first met in 1970, and was able to agree in principle to a list of items that, when exported, would "trigger" the imposition of safeguards in accordance with Article III(2)(b) of the NPT. Published by the IAEA in September 1974, the list is known as INFCIRC/209. The NSG, or the London Club, was organised by the US during the Ford administration in an attempt to buttress the NPT by controlling, if not preventing, the export of certain sensitive nuclear items. To this end, the group secretly agreed on voluntary guidelines for the transfer of nuclear technology, which were subsequently communicated to the IAEA and published in INFCIRC/254.[4] These guidelines are a set of identical but unilaterally adopted voluntary undertakings to use restraint in the transfer or retransfer of sensitive nuclear technology, equipment, and materials by implementing certain conditions. These conditions include assurances of non-explosive use by the recipient, effective physical protection for nuclear materials, IAEA safeguards agreements, as well as other measures also intended to limit the spread of nuclear weapons. The approach of Suppliers towards implementing these conditions differs. They include requiring comprehensive safeguards (safeguards on all nuclear materials and facilities within a country) as a condition of significant exports, prior consent provisions for reprocessing or plutonium use, agreements on handling dual-use items and technologies, or an absolute prohibition on the export of enrichment or reprocessing technologies and equipment.[5] This reflects the difference in interests among Supplier States, regardless of their commitment to the issue of non-proliferation of nuclear weapons.

In this chapter, the following shall be discussed:

- The evolution of international nuclear commerce.
- The role of supplier groups in controlling the transfer of nuclear material and equipment to NNWS.
- The role of emerging suppliers and the future of the supplier regime.

2. *Evolution of International Nuclear Commerce*

During the 1950s and 1960s, the US occupied a dominant position in the global nuclear market. Prior to 1963, this market was limited primarily to research reactors, most of which were supplied by the US, along with

[4] Ibid, pp. 39–41.

[5] Ibid, p. 41.

training services and the enriched uranium needed to fuel them. Large-scale trade in nuclear power facilities did not begin until after 1963, when General Electric, followed by Westinghouse, made available turn-key light-water reactors (LWR) at prices that were competitive with fossil-fuel plants. These US vendors soon acquired a dominant international market position for their reactors. US enrichment services also enjoyed a virtual monopoly in the supply of uranium fuel for the LWR. During this period, US firms engaged in a number of joint ventures and technology licensing agreements with foreign companies. The US nuclear industry also initially found the EURATOM to be a "convenient channel" for European sales. By the mid-1970s, however, some of these foreign firms, through heavy government subsidies, emerged as US rivals. Canada, France, West Germany, and the Soviet Union all supported companies that offered real competition in the international reactor market. The US monopoly in uranium enrichment was also diminished by the start up of Eurodif (a French joint venture with Belgium, Iran, Italy, and Spain) and Urenco (a British-Dutch-West German consortium) and by Soviet suppliers. European firms in Belgium, Italy, Sweden, and Switzerland also established a niche for themselves as providers of nuclear technology components and services.[6]

One consequence of the oil crisis of 1973–1974 was a sharp rise in the demand for nuclear energy forecasts for the rest of the century. Opportunities for nuclear exports also appeared to shift from the industrialised world to developing regions, where support for the non-proliferation regime was often problematic. The developing world emerged as an attractive market to non-US suppliers among which the US nuclear industry was not yet entrenched. France and West Germany, in particular, were ready to supply nuclear technologies that the US had withheld because of their proliferation implications. Also, international nuclear relations in the mid-1970s were affected by the Indian nuclear detonation in May 1974. The Indian test demonstrated that even a relatively poor, developing state could, with sufficient political will, circumvent the proliferation barriers imposed by the NPT. At the time, the non-proliferation proponents were troubled by the effect that the Indian explosion might have on the many countries that had initiated or accelerated nuclear power programs in the aftermath of the 1973–1974 oil crises.[7]

---

[6] Ibid, pp. 3–4.
[7] Ibid, p. 5.

Unlike most other commodities, products for the nuclear industry do not benefit from international mechanisms helping the distribution of goods. Neither is there a coherent set of rules or general agreements governing the different economic aspects of nuclear trade. Instead, its regulation takes many forms: instruments of competent international organisations, commitments undertaken in bilateral or multilateral agreements, inter-governmental arrangements whose legal status is somewhat uncertain, and resolutions with an even hazier scope adopted in international conferences and—not to be overlooked—national decisions of a political nature. Most of the texts record mutual obligations and establish controls ensuring that the goods exchanged have an exclusively peaceful vocation. This legal structure is capped by the Safeguards System of the IAEA which, in international law, is a unique example of acceptance of direct inspections on the territory of sovereign states.[8]

The evolution of nuclear activities in the past thirty years brought with it a corresponding growth in nuclear trade.[9] However, the examination of the law, applicable to international transfers of nuclear material and equipment, reveals that the goal of preventing nuclear-weapon proliferation generally prevails over promoting economic relations in the area of nuclear commerce.[10]

Practice proved that preventing nuclear weapons proliferation not only prevailed over promoting economic relations, but it also prevailed over promoting international cooperation for the peaceful uses of nuclear technology. The threat of the dual use of nuclear technology prevented nuclear suppliers from taking the lead in transferring nuclear technology to NNWS. This has affected the use of nuclear technology for peaceful purposes around the world, especially in the field of nuclear energy, which remained at its highest in the territories of nuclear Suppliers and at its lowest in the rest of the world.

---

[8] Nuclear Energy Agency, 'The Regulation of Nuclear Trade: Non-Proliferation—Supply-Safety: International Aspects,' Volume 2, 1988, p. 9.

[9] 'Nuclear Power in the World Today,' Nuclear Issues Briefing, Paper 7, June 2003. "In 1987, 417 reactors produced more than 16% of the world's electricity, exceeding 23 per cent in the OECD area and 30% in the Common Market, in 2002, the percentage of nuclear power share in world electricity remained the same, nuclear power provided over 16% of the world's electricity, almost 24% of electricity in OECD countries, and 35% in the EU. In 2003, there were 439 commercial nuclear reactors operating in 31 countries, with some 360,000 MWe of total capacity."

[10] Nuclear Energy Agency, 'The Regulation of Nuclear Trade: Non-Proliferation—Supply-Safety: International Aspects,' Volume 2, 1988, p. 9.

3. *International Associations Controlling the Movement of Nuclear Supplies*

Concerned with the threat of nuclear proliferation, suppliers formed organisations and committees to restrict the movement of nuclear material, equipment and technology around the world. These forums played a substantial role in controlling, and sometimes restricting, the movement of nuclear material and equipment to NNWS.

The role of these players in the field of nuclear trade will be discussed as follows:

1. Nuclear Suppliers Groups:
    i. Nuclear Exporter Committee.
    ii. The Nuclear Supplier Group—the London Club (NSG).
2. Assessment of the legal nature of the commitments and effects of establishing the Nuclear Exporters' Committee and the NSG.
3. The effects of establishing the NSG and the Zangger Committee.
4. Other elements in the regulation of nuclear trade.

3.1. *Nuclear Suppliers Groups*

Nuclear Exporters Committee (Zangger Committee)

Framework for Negotiations

The Zangger Committee was formed by supplier countries to decide how to interpret the rather generally drafted export control obligation in the NPT. They determined a list of goods, called the "Trigger List", whose export would "trigger" safeguards, and defined procedures and conditions under which nuclear exports would be licensed. These Zangger Committee "Understandings", ready in 1972, were published in September 1974 as IAEA document INFCIRC/209. The Trigger List has been amended several times since then. Subsequently, new member States communicated letters to the DG of the IAEA with the information that they have joined the NSG.[11]

[11] Schmidt, Fritz, 'Zangger Committee,' paper presented before the 2nd Seminar on Export Controls in Nuclear Non-Proliferation held at the UN, New York, 8–9 April 1999, p. 25.

Declarations of Policy
In each of the above-mentioned letters, the Government of the country in question indicated that it had been called upon to examine procedures in relation to exports:

- of source or special fissionable material,
- of equipment and material especially designed or prepared for the processing, use or production of special fissionable material, and
- committing itself not to provide any source or special fissionable material unless it is subject to safeguards under an agreement with the IAEA.[12]

The letters conclude by saying that the Governments in question have decided to act in accordance with the provisions of the memoranda attached to the declaration of policy.

Memoranda to the Declaration of Policy
Memorandum A. Memorandum A deals with source and special fissionable material as defined in Article XX of the IAEA Statute. Memorandum A states that "the Government is solely concerned with ensuring the application of safeguards in NNWS not party to the NPT with a view to preventing diversion of the safeguarded nuclear material from peaceful purposes to nuclear weapons or other nuclear explosive devices." Consequently, each government undertakes to specify to the recipient State, as a condition of supply, that the source or special fissionable material, or special fissionable material produced in or by the use thereof, shall not be diverted to nuclear weapons or other nuclear explosive devices. It also undertakes to ensure that IAEA safeguards will be applied to the source or special fissionable material in question, when the material reaches its destination, at the very latest, and that said safeguards will apply in the event of re-exporting (re-transferring) the material to an NNWS not party to the NPT.[13]

Memorandum B. Memorandum B deals with equipment or material especially designed or prepared for the processing, using or producing of special fissionable material, and includes:[14]

[12] Nuclear Energy Agency, 'The Regulation of Nuclear Trade: Non-Proliferation—Supply-Safety: International Aspects,' Volume 2, 1988, p. 77.
[13] Ibid, p. 78.
[14] Ibid.

- "Trigger List" of equipment, components and materials necessary for the nuclear fuel cycle, including reactors. Non-nuclear materials are also included, i.e. deuterium and heavy water, graphite and reprocessing plants, plants for the fabrication of fuel elements and equipment especially designed or prepared for the separation of isotopes of uranium. There was, however, no ban on transfers of sensitive technologies used in enrichment and reprocessing. Moreover, safeguards were required only for the transferred materials and equipment.
- Implementation of the main principles for granting export and retransfer authorisations, which are similar to those mentioned above under Memorandum A. (safeguards, non-diversion to military uses … etc).

Each government reserved the right to interpret and implement its commitment under these arrangements with regard to applying safeguard requirements. The main result of this agreement within the Zangger Committee was to subject all future nuclear transactions—whether or not within the NPT—to a limited but more stringent control system.[15]

Resumption of Zangger Committee Activities after 1978

After a period during which co-ordination of nuclear export policies were effected within the framework of the London Club, the countries participating in the Zangger Committee work resumed consultations. Memorandum B was clarified and completed by identical letters dated 1st September 1978 and addressed to the DG of the IAEA and were distributed by the Agency to all its Member States (INFCIRC/209/Mod. 1, December 1978). In these letters, Memorandum B was expanded to include zirconium tubes, plants for the production of heavy water, deuterium and deuterium compounds and certain equipment used for isotope separation plants. Letters were sent again to the DG, as of January 1984, developing and clarifying the Trigger List with regard to assemblies and components especially designed or prepared for use in isotope separation facilities using the gas centrifuge enrichment process; (INFCIRC/209/Mod. 2, February 1984). On 1st July 1985, the Annex to a new series of letters clarified provisions concerning equipment performing essential tasks in reprocessing plants (INFCIRC/ 209/Mod.3).[16]

---

[15] Ibid, p. 17.
[16] Ibid, p. 78.

The new restrictions on components designed to be used in centrifuge enrichment process are against the spirit of the NPT. The IAEA/NPT rules do not restrict such transfers as long as they fall under IAEA safeguards and are used solely for peaceful purposes. These restrictions go beyond the requirements of the NPT.

NSG—London Club

Framework for Negotiations

In response to various violations of the NPT export guidelines by several developed countries, the US initiated a second nuclear suppliers' group. In 1975, the US, Canada, France, West Germany, Japan, the UK, the US, and the USSR met in London to establish the NSG, which was subsequently expanded to forty-five members including various importing countries. The negotiations aimed at achieving a consensus among exporting countries on a more restrictive export policy than that found in the NPT, and which would require full-scope safeguards for all nuclear activities in importing NNWS and would call for an embargo on the most sensitive parts of the fuel cycle, i.e. reprocessing and enrichment. The consultations tried to harmonise export policies from the safeguards and control angle for transfers of "nuclear items" outside the framework of the IAEA and the NPT; this also applied to technology transfers, control of retransfers and physical protection.[17]

On 11 January 1978, the involved States, and other States that joined the group afterwards, sent a letter to the DG of the IAEA to inform him of their decision to act in accordance with the principles stipulated in an attached document which dealt with the export of nuclear material, equipment or technology, asking him to inform all Member States of this fact (INFCIRC/254, February 1978, "Guidelines for the Export of Nuclear Material, Equipment or Technology." The NSG urged for the adoption of domestic laws requiring that exports of any "trigger list" items had to be licensed.[18]

---

[17] Ibid, p. 79. "Regarding the second point, the stricter proposals (subsequently adopted in the United States legislation) did not achieve consensus; the compromise solution was to recommend particular caution for the export of 'sensitive' materials (uranium enriched to more than 20 per cent and plutonium) and equipment." See also Nuclear Energy Agency, 'The Regulation of Nuclear Trade: Non-Proliferation—Supply-Safety: International Aspects,' Volume 2, 1988, p. 18.

[18] IAEA, INFCIRC/254, 'Communication Received from Certain Member States Regarding Guidelines for the Export of Nuclear Material, Equipment or Technology,' February 1978.

Criticism of the guidelines established by the Club focuses on the issue of restricting NNWS access to sensitive nuclear technology for peaceful purposes beyond the IAEA/NPT restrictions, as the full scope safeguards imply that the Supplier State, when exporting any item on the trigger list, will demand safeguards on all the nuclear facilities in the recipient State and not only on the exported item. In addition, the Dual-Use Guidelines state that "suppliers should not authorise transfers of equipment, material, or related technology ... when there is an unacceptable risk of diversion [to a nuclear explosive activity or an unsafeguarded nuclear fuel cycle activity] or when the transfers are contrary to the objective of averting the proliferation of nuclear weapons." Hence, States whose "*bona fides* are *questioned* by *one* or more *nuclear supplier(s)*" will be denied exports from one of the member States of the NSG.[19] For these reason, the judgments on nuclear and dual-use exports are always subjective, not objective.[20]

Declarations of Policy and Reservations

The State of the London Club exchanged a series of letters in which they endorsed a uniform code of conduct for international nuclear exports. The major provisions of the code, which is essentially a gentleman's agreement rather than a treaty, require that before sensitive nuclear materials, equipment, or technology are transferred, the recipient state must:[21]

- Pledge not to use the transferred materials, equipment, or technology in the manufacture of nuclear explosives;
- Accept, with no provision for termination, international safeguards on all transferred equipment or technology, including any facility that replicates or otherwise employs transferred technology;
- Provide adequate physical security for transferred nuclear facilities and materials to prevent theft and sabotage; and

19 Nasseri, Cyrus, 'A Perception for Evolution: the NSG's Impact on Non-proliferation and the Right to Access,' paper presented before the 2nd Seminar on Export Controls in Nuclear Non-Proliferation held at the UN, New York, 8–9 April 1999, p. 54.

20 Stratford, Richard J.K., 'The Practice of Export Controls: Effects on Trade and Peaceful Nuclear Activities,' paper presented before the 2nd Seminar on Export Controls in Nuclear Non-Proliferation held at the UN, New York, 8–9 April 1999, p. 39.

21 Potter, William C., 'International Trade and Non-Proliferation: The Challenge of the Emerging Suppliers,' 1990, p. 5.

- Agree not to re-transfer the materials, equipment, or technology to third countries unless they too accept the constraints on use, replication, security, and transfer, and unless the original supplier concurs in the transactions.[22]

It was agreed that any changes of the Guidelines require unanimous consent, especially those changes which might arise from a reconsideration of common safeguards requirements. However, exceptions made to safeguards duration and coverage provisions only require prior consultation with the other parties, but some States expressed reservations (allowing governments to apply a wide interpretation, requirement or changing domestic legislation … etc) or made known their intention to impose additional requirements.[23]

The Member States of the European Community with the exception of France specified that they would apply the Guidelines to intra-Community trade in the light of commitments undertaken by virtue of the EURATOM Treaty. Furthermore, the USSR and certain socialist countries expressed their intention to make the supply to an NNWS of the items mentioned in the Trigger List conditional upon the application of IAEA safeguards to all nuclear activities within the State in question. For its part, Canada reserved the right to act in accordance with "other principles considered pertinent by it." Similarly, Australia reserved the right to take certain other requirements into account. The US adopted a different approach and did not introduce any reservation in its letters, but in fact subsequently introduced in its legislation (NNPA) stricter requirements than those in the Guidelines.[24]

Documents Attached to the Declarations of Policy

There are three documents attached to the letters sent by the above-mentioned States:[25]

- an *Appendix* entitled "Guidelines for Nuclear Transfers";
- *Annex A* entitled "Trigger List referred to in Guidelines", to which is attached a separate Annex entitled "Clarifications of Items on the Trigger List";

[22] Ibid, pp. 5–6.

[23] Nuclear Energy Agency, 'The Regulation of Nuclear Trade: Non-Proliferation—Supply—Safety: National Regulations,' Volume 1, 1988, p. 79.

[24] Ibid.

[25] Ibid, pp. 79–80.

- *Annex B* entitled "Criteria for Levels of Physical Protection", to which is attached a Table categorising nuclear material.

The purpose of the *Guidelines* is to lay down the fundamental principles governing the safeguards and controls that must be applied to any export of nuclear items to NNWS. They are based on the Trigger List drawn up by the supplier States which took part in the work of the NSG, as well as on common criteria, agreed between these States, relating to technology transfers. In addition, they define application procedures for certain special cases and provide for "supporting activities."[26]

"Fundamental Principles" of General Application

The fundamental principles stated in the guidelines can be summarised as follows:[27]

a. Suppliers should authorise transfer of items identified in the Trigger List only upon formal governmental assurances from recipients which explicitly exclude uses that could lead to the creation of any nuclear explosive device.
b. Suppliers should concern themselves with physical protection and require that all nuclear materials and facilities identified in the Trigger List be placed under effective physical protection to prevent unauthorised use and "handling."
c. Suppliers should transfer Trigger List items only when these are covered by IAEA safeguards, with duration and coverage provisions in accordance with the recommendations already included in a Memorandum from the DG of the IAEA to the BG (GOV/1621 of 20th August 1973) aiming at encouraging more standardised safeguards agreements—particularly with respect to their duration and termination. Hence, in order to harmonise the duration and termination clauses of safeguards agreed within the framework of the pre-NPT IAEA system (INFCIRC/66/Rev.2), the Memorandum provides:[28]
   - that the duration of the safeguards agreement should be related to the period of actual use of the supplied items in the recipient State; and
   - that the provisions for terminating the agreement should be formulated in such a way that the rights and obligations of the

[26] Ibid, p. 80.
[27] Ibid.
[28] Ibid.

parties continue to apply in connection with supplied nuclear material and with special fissionable material obtained, produced, processed or used in or in connection with supplied nuclear material, equipment, facilities or non-nuclear material (or obtained through such items); until such time as the Agency has terminated the application of safeguards thereto, in accordance with its Safeguards System (return to supplier country or transfer under safeguards to a third country; no longer usable for any nuclear activity).

- Exceptions should be allowed only after consultations between the parties.

In conclusion, the only way to introduce an integral safeguards system as a common requirement is through a joint reconsideration of the Guidelines.[29]

Although the Guidelines adopted by the Group constituted a compromise between two different extremes (i.e., banning the transfer of sensitive nuclear technology vs. permitting such transfers with no more restrictions than required by the NPT), the fact remains that the Group implemented many additional restrictions over the years on the transfer of nuclear technology to NNWS. The US, as previously mentioned, imposed numerous restrictions by means of legislation on export of nuclear material and equipment. Recent developments have revealed that consecutive American administrations have persisted in seeking a ban on the transfer of certain technologies to NNWS, whether these technologies will be under safeguards or not, as with its current efforts to obstruct any supply of uranium-enrichment technology to Iran. The question at hand would be does US suspect that the current rules are not adequate to prevent the non-diversion of transferred technology? Or are these efforts directed towards ensuring the monopoly and dominance of certain players over the nuclear market?

Special Controls on Sensitive Exports (paragraph 7). This paragraph concerns the transfer of sensitive materials, facilities, technology and weapons-usable materials. Despite the absence of a precise definition of these terms it can, however, be deduced from paragraph 7 that they include, facilities, equipment or technology for enrichment or reprocessing. As for materials with a military capacity either, because of their nature or

[29] Ibid.

degree of enrichment, they too are not explicitly defined; but the context (in particular paragraph 8 of the Appendix and the categorisation of materials for physical protection purposes) indicates with sufficient clarity that this concerns unirradiated plutonium (except that with a high Pu 238 content) and uranium enriched to 20 per cent or more. Although paragraph 7 does not explicitly refer to heavy-water production facilities, it would seem that such facilities should also be considered as sensitive exports for application of controls as provided by this paragraph. "Sensitive" technologies are those referred to in paragraph 6 of the Guidelines and defined in Annex A, Part B, paragraph 1 thereof (Trigger List). Although heavy water as such does not constitute a major component in the production of nuclear weapons or nuclear explosive devices, it is still considered part of the Trigger List items.[30] This is an indication that the Group works in a manner closer to banning than to facilitating the transfer of nuclear technology for peaceful purposes.

As for transfers related to enrichment and reprocessing, the basic principle is that suppliers should limit such transfers and that if, nevertheless, they are to take place, suppliers should encourage recipients to accept, as an alternative to the construction of plants on a strictly national basis, supplier involvement and/or other appropriate multinational participation in resulting facilities. Suppliers should also promote projects for multinational fuel cycle centres.[31]

Special Controls on the Export of Enrichment Facilities, Equipment and Technology (Paragraph 8). Under the Guidelines, the recipient nation must agree that neither the transferred facility, nor any facility using the technology involved, will be designed or operated for the production of uranium enriched to more than 20% enriched uranium without the consent of the supplier nation, and without notifying the IAEA.[32]

Controls on Supplied or Derived Weapons-Usable Material (Paragraph 9). Suppliers recognise the importance of including in agreements on supply of nuclear materials, or of facilities which produce weapons usable material, provisions "calling for" mutual agreement between the supplier and recipient on arrangements for reprocessing, storage, alteration, use, transfer or re-transfer of any weapons-usable material involved.

---

30 Ibid, p. 81.

31 Ibid.

32 Ibid.

This complex phrasing in effect concerns "prior consent" to reprocessing which certain countries wished to impose when supplying reactors or their fuels. However, since no consensus was reached in this respect, the last sentence of paragraph 9 simply provides that suppliers "should endeavour to include such provisions whenever appropriate and practicable."[33]

Controls on Re-transfer (Paragraph 10). Suppliers should transfer Trigger List items, including technology related to reprocessing, enrichment, or heavy-water production facilities, only upon assurance that in the case of[34] re-transfer of such items or transfer of Trigger List items derived from facilities, or with the help of equipment or technology originally transferred by the supplier;

- the recipient will have provided the same assurances as those required by the supplier for the original transfer.

In addition, the supplier's consent should be required for:

- any retransfer of the facilities, "major critical components" or technology relating to reprocessing, enrichment, or heavy-water production facilities;
- any transfer of facilities or "major critical components" derived from those items;
- any re-transfer of heavy water or weapons-usable material.

Non-Proliferation Principle (Paragraph 11). Suppliers should authorise transfer of items or related technology identified in the Trigger list only when they are satisfied that the transfers would not contribute to the proliferation of nuclear weapons or other nuclear explosive devices.[35]

Supporting Activities

A primary objective of suppliers is to promote international cooperation concerning exchanges of "physical security" information, a term which seems, given the context, to concern physical protection, and to indicate measures intended to make it more difficult to use nuclear materials

---

[33] Cassidy, Stephen, 'The Newest Member of the Nuclear Club: Pakistan's Drive for a Nuclear Weapons Capability and US Nuclear Non-Proliferation Policy,' 1988/1989.

[34] Nuclear Energy Agency, 'The Regulation of Nuclear Trade: Non-Proliferation—Supply—Safety: National Regulations,' Volume 1, 1988, p. 81.

[35] Stratford, Richard, 'The Practice of Export Controls: Effects on Trade and Peaceful Nuclear Activities,' 1999, p. 38.

for explosive purposes; such exchanges cover both the protection of nuclear materials in transit, and recovery of stolen nuclear materials and equipment.[36]

Furthermore, each Supplier State is obliged to encourage the "designers" and makers of sensitive equipment to construct it in such a way as to facilitate the application of safeguards.[37]

When considering transfers, suppliers are also asked to exercise prudence, and to take into consideration all the circumstances of each case. They are, in particular, requested to assess whether there is any risk that technology transfers or re-transfers of other than sensitive technology (reprocessing, enrichment, or heavy-water production), might result in the production of unsafeguarded nuclear materials. Lastly, suppliers should consult through regular channels, either amongst themselves on questions relating to the implementation of the Guidelines, or with the Governments concerned, specific cases which, due to a risk of conflict or international disorders, are particularly sensitive.[38]

In the event of a breach of supplier-recipient agreements relating to the implementation of the Guidelines—particularly in the case of an explosion of a nuclear device, or illegal termination or violation of IAEA safeguards by a recipient, suppliers should promptly be consulted through diplomatic channels in order to determine and assess the reality and extent of the alleged breach, and agree, in the spirit of Article XII of the IAEA Statute concerning safeguards, on an appropriate response. Such a response could include termination of nuclear transfers to the recipient in question.[39]

Trigger List and Clarifications of Items on the List

Annex A of the Guidelines constitutes the "Trigger List" referred to therein. An Annex to Annex A in turn clarifies certain items on the Trigger List. It is divided into two parts:

Part A deals with material and equipment. Source or special fissionable material is defined as in Article XX of the IAEA Statute. However, "small quantities" exported to a given country within a period of twelve months are exempted from application of the Guidelines.[40]

---

[36] Nuclear Energy Agency, 'The Regulation of Nuclear Trade: Non-Proliferation—Supply—Safety: National Regulations,' Volume 1, 1988, p. 81.

[37] Ibid.

[38] Ibid, p. 82.

[39] Ibid.

[40] Ibid.

The first Trigger List of 1974 contained nuclear material, source material, special fissionable material, reactors and equipment there for, reactor fuel charging and discharging machines, reactor control rods, reactor pressure tubes, zirconium tubes, primary coolant pumps, non-nuclear materials for reactors, deuterium and heavy water, nuclear grade graphite, plants for the reprocessing of irradiated fuel elements, and equipment, especially designed or prepared there for, plants for the fabrication of fuel elements, equipment, other than analytical instruments, especially designed or prepared for the separation of isotopes of uranium. This list has been updated regularly since 1974.[41]

Part B of the Trigger List (Annex A) sets the common criteria for the transfer of sensitive technology. It may be recalled that paragraph 6 of the Guidelines concerned with making the transfer of technology that is related to facilities for reprocessing, enrichment or heavy-water production subject to conditions concerning prohibition of obtaining nuclear explosive devices, or physical protection measures and the application of safeguards, emphasises the need to extend these obligations to "major critical components," to technology resulting from transferred facilities, and to facilities "of the same type" constructed in the recipient country during an agreed period.[42]

The purpose of Part B is to define the concepts underlying the requirements laid down by paragraph 6 of the Guidelines. To this end, Part B specifies the meaning to be given to the word "technology" and the terms "major critical components" and facilities "of the same type," thus making clear that these definitions should be interpreted differently from those applicable to Part A. In addition, criteria are laid down which give a fairly wide interpretation of the scope of paragraph 6 in two situations which might give rise to doubt:[43]

- firstly, by specifying that a transfer in the aggregate of a significant fraction of the items essential for the operation of a facility referred to in paragraph 6, together with the know-how for construction and operation of that facility, shall be deemed to be a transfer within the meaning of paragraph 6, even though no major critical component is included in the transfer; and secondly,

[41] Schmidt, Fritz, 'Zangger Committee,' 1999, p. 26.

[42] Nuclear Energy Agency, 'The Regulation of Nuclear Trade: Non-Proliferation—Supply—Safety: National Regulations,' Volume 1, 1988, p. 82.

[43] Ibid.

- by specifying that a similar approach would be applied to define facilities "of the same type" and "major critical components" in cases where methods, different from those based on physical or chemical processes already defined as "of the same type" for the design, construction and operation of reprocessing, enrichment or heavy-water production facilities.

Furthermore, Part B lays down a minimum period of twenty years, from the date of the commissioning of the facility in question, during which IAEA safeguards must be applied to facilities in the recipient country the same type as the transferred facility.[44]

The Annex to Annex A contains clarifications of items on the Trigger List. The purpose of this Annex is essentially to identify the parts which make up all the major components of a reactor, pressure vessel, reprocessing plant, etc. As already indicated, the Guidelines apply as a rule to the whole set of these items, with Governments reserving to themselves the right to apply the procedures to an item delivered separately.[45]

Annex B

Annex B, which constitutes the last part of the Guidelines, lays down the criteria for levels of physical protection.[46]

3.2. *Assessment of the Legal Nature of the Commitments undertaken Pursuant to the Work of the Zangger Committee and the London Club*

The legal scope of these commitments has been the subject of much conjecture. Do they raise obligations under international law between the States having undertaken them and between such States and those IAEA Member States which have not? Furthermore, if it is felt that the Guidelines do raise obligations, the precise nature and extent of such obligations should be clarified.[47]

Within the work of the Zangger Committee the word "commitment" has been adopted in the absence of a more neutral term, and should not be taken as prejudging the nature and legal effects of the undertakings

[44] Ibid.
[45] Ibid.
[46] Ibid.
[47] Ibid.

involved. Some writers have described the NSG Guidelines as a "gentlemen's agreement", or likened them to what has been termed "international soft law".[48]

A "gentlemen's agreement" should of course not be taken in this case to mean a personal arrangement by negotiators which is not legally binding to the parties represented, which is its original meaning because, the term has evolved to include an arrangement between States which does create obligations. However these obligations, if not performed, do not involve the international liability of the State, inasmuch as no legal proceedings can be brought before the International Court of Justice (ICJ). Neither is a "gentleman's agreement" restricted, in modern times, to declarations of general policy. Certain such agreements interpret or fill in gaps in treaties or the statutes of international organisations; others define the regulatory framework in which relations between States should be established. There are sound arguments in doctrine for defining the London Club and Zangger Committee Guidelines in this way.[49]

Despite the confidential nature of discussions, especially those of the London Club, there is reason to suppose that, as early as their 1975 meeting, the "founding" States agreed not to conclude a formal international treaty, which meant that neither the London Club Guidelines nor the Zangger Committee Memoranda needed to be registered with the UN Secretariat (Article 102 of the Charter). It nevertheless remains true that concerted action was taken—a meeting of minds after prior negotiation—as shown not only by the identical or similar nature of the letters addressed to the DG of the IAEA, but also by the regulatory nature of the Annexes and the decision not to amend the London Club Guidelines except by unanimous consent, and to continue to hold consultations on certain questions of implementation. The Zangger Committee practice shows even more strongly a desire to round out the initial agreement (i.e., the NPT).[50]

Lastly, if it is accepted that "gentlemen's agreements" are governed by international law, but not the part of that law dealing with international treaties, they must be performed in good faith. Thus, a State which has undertaken a commitment within the framework of a gentleman's agreement is not entitled arbitrarily or secretly to depart from or change it. This is true both with regard to the other States committed under the

---

48 Ibid.
49 Ibid.
50 Ibid.

agreement as well as, given the public request to the IAEA, to the other Member States of the Agency. However, it is accepted that a participating State is free to decide to no longer participate in the consultations or to undertake new commitments instead.[51]

That the commitments published by the IAEA under the reference INFCIRC/209 or 254 are not purely potestative also follows from the other possible interpretation, which is supported by various countries—that of unilateral undertakings. The ICJ today recognises that unilateral undertakings by a State, expressed publicly and with the intention of commitment, have, even if outside the framework of international negotiations, a binding effect. The State concerned is then obliged, by virtue of the principle of good faith, to follow a line of conduct in accordance with this declaration. It follows that while a declaration by a State limiting its freedom of future action must certainly be interpreted restrictively, it does not carry with it an arbitrary power of review. According to some, a State could terminate its undertakings by means of unilateral action, in the same manner as it would relieve itself of obligations derived from a treaty. Thus a fundamental change of circumstances, which may be likened to the *rebus sic stantibus* of the law of treaties, could justify a repeal or a review of the obligations entered into.[52]

Following this line of thought, a distinction should perhaps be made between two situations:[53]

- that of those States which participated in the consultations undertaken within the framework of the Zangger Committee or London Club; in this case, the relations among these States would be governed by a gentlemen's agreement, whereas those between any such State and States not having taken part in these consultations would be governed by the rules on unilateral undertakings;
- that of the many other States which, subsequently, asked the IAEA to record and publish their decision to act in compliance with the prin ciples laid down in the Zangger Committee or London Club documents. Such a decision would count as a unilateral undertaking with regard both to countries having participated in these consultations as well as to third States. An alternative interpretation would be to consider the commitment undertaken by these States with regard to

[51] Ibid.
[52] Ibid, p. 83.
[53] Ibid.

countries having participated in the consultations as an adhesion to the gentlemen's agreement, subject, wherever necessary, to the relevant provisions of the law of treaties relating to adhesion. Given the present state of the law, however, this latter interpretation appears somewhat unorthodox.

In conclusion, it may be noted that there is an essential difference between the Zangger Committee and the London Club. The declarations of the former give a common interpretation of an Article of the NPT which involves a specific undertaking by the contracting parties. On the other hand, the London Club Guidelines do not refer to a formal provision and simply express the decision of participating countries to act in accordance with principles defined in the Annex. Therefore, the two initiatives cannot be placed on the same level.[54]

4. *Effects of establishing the NSG and the Zangger Committee*

The NSG Agreement of 1976 does not ban nuclear transfers to non-parties of the NPT or to states that refuse to accept "full-scope safeguards" (that is, to place all of their facilities under IAEA safeguards). Nor are members of the group legally bound to act according to the Trigger list guidelines. The list is simply a multilateral statement of national policy. As such, it seeks to prevent the erosion of safeguards by unregulated competition. However, the assertion of authority over nuclear trade by the traditional suppliers in the 1970s was achieved with little regard for the concerns of the importing states. Developing countries, in particular, tended to perceive both the Zangger Committee and the NSG as instruments of the industrialised States which sought to establish a nuclear cartel for the purpose of assuring the economic dependency of Third World nations.[55] The same discontent still exists today.

Defending the NSG position before the 1995 Review and the Extension Conference of the States Parties to the NPT, the Australian representative stated that "A few states have sought to portray export controls as a North-South issue, as involving a cartel or as a conspiracy which goes beyond the legitimate terms of the NPT. This is not so. Export licensing

[54] Ibid, p. 84.

[55] Potter, William C., 'International Trade and Non-Proliferation: The Challenge of the Emerging Suppliers,' 1990, p. 6.

arrangements do not impede legitimate nuclear trade and cooperation. Rather, they are an important part of the environment of long-term assurance and stability that underpins nuclear cooperation. For NPT parties abiding by their Treaty obligations—and that is clearly the vast majority of countries—such controls do not constitute any sort of impediment. Nor do informal arrangements such as the NSG operate as any sort of cartel."[56]

In 1999, Mr. Alec Baer supported this argument. He stated that "In spite of all the criticism that it was a cartel aiming at preventing the spread of nuclear technology to non-members, the NSG has definitely been a positive factor of non-proliferation. Its role and importance have clearly not been limited to formal denials of items from one or the other list. A not quantifiable, informal influence of the NSG on the policies of various States is not to be denied, be it through the use of the traditional diplomatic channels or through other contacts between Member States and other States. It has been said that the whole NSG scheme can never prevent a state from proliferating, but will only slow it down. This may be true, but gaining time is also a profitable approach to such problems."[57]

The guidelines for nuclear transfer worked out by the London suppliers' club are a vivid proof that constraints and limitations imposed on nuclear cooperation do in fact exist. Although the objectives pursued by the "club" are quite legitimate, i.e., the prevention of the proliferation of nuclear weapons and the elimination of commercial competition as a factor in negotiating safeguards, the guidelines may well lead some Supplier States to follow the US example by refraining from the transfer of basic technology, especially in the sensitive areas of uranium enrichment, reprocessing and heavy water production. Moreover, in the process of restricting the transfer of technologies which have potential military nuclear applications, the guidelines may have the impact of restricting as well the flow of basic knowledge in areas not necessarily confined to the nuclear field. In fact, the spirit which inspired the formulation of the Guidelines is very much in favour of denying certain technologies to developing countries regardless of the principles enshrined in Article IV of the NPT. The mere fact that the guidelines were worked out in secrecy

[56] Evans, Senator Gareth, the former Foreign Minister of Australia, 'Statement on 18 April 1995 to the 1995 Review and Extension Conference of the States Parties to the Treaty on Non-Proliferation of Nuclear Weapons,' 1995.

[57] Koplow, David, 'The Jurisprudence of Non-Proliferation: Taking International Law Seriously,' 1992, p. 10.

and without the participation of the developing countries, and then were undisclosed for a long period of time before their release, is also a violation of the spirit of cooperation prescribed by Article IV.[58]

The paradox here does not lie in the formulation of NPT rules, but in the spirit by which NSG nuclear suppliers act upon them. If those suppliers had faith in the NPT/IAEA system, they would not have applied further restrictions on NNWS. They insist on full-scope safeguards. However, under the IAEA/NPT rules only the material or equipment the supplier provides should be placed under IAEA safeguards. Why would the Supplier State impose on the recipient State further obligations than those required by the NPT? This breaks the principles of sovereignty and equality between states, thus exposing an imbalance of power.

The 2000 Review Conference of the NPT discussed the measures taken by the supplier groups and requested that their guidelines not hamper the transfer of nuclear technology for peaceful purposes. It requested that any supplier arrangement should be transparent and should continue to take appropriate measures to ensure that the export guidelines of the Suppliers do not hamper the development of nuclear energy for peaceful uses by States parties, in conformity with Articles I, II, III and IV of the NPT. The Conference also recommended that transparency in export controls should continue to be promoted within a framework of dialogue and cooperation among all interested States party to the Treaty.[59] NNWS concerns raised in the NPT Review Conferences revealed that the implementation of the Guidelines of the NSG and the lack of transparency in its work hampered the development of nuclear technology for peaceful purposes in NNWS, especially those in the developing world.

5. *Other Elements in the Regulation of Nuclear Trade*

5.1. *Declaration of Common Policy by the European Community*

On 20th November 1984, the Foreign Ministers of the ten Member States of the European Community adopted a declaration of common policy. They noted that "all the Member States of the European Community

[58] Shaker, Mohamed, 'The Nuclear Non-Proliferation Treaty: Origin and Implementation 1959–1979,' 1980, pp. 332–333.

[59] NPT Review Conference, NPT/CONF.2000/28, 'Final Document of the 2000 Review Conference of the Parties to the NPT,' 24 May 2000, p. 9.

had adopted, as unilateral undertakings, the London Club Guidelines, published as INFCIRC/254, a common, fundamental set of rules for all the Member States in relation to their nuclear exports." They also stated that transfers of nuclear material, equipment and technology could be made without restriction between the Member States in a manner compatible with the requirements of both non-proliferation and the free movement of goods, on condition that the provisions of the Treaty of Rome and the competence of the Member States were respected, subject to the following additional arrangements:[60]

- until such time as they are used, separated plutonium and uranium enriched to more than 20 per cent will be stored by the Member States at the place of separation or enrichment, or at the places of fabrication of fuels containing such materials, or in a store established and administered by a Member State, or in a place to be determined by common agreement between the Member States concerned;
- such plutonium and uranium will be transferred by the Member States upon receipt of a certificate from the consignee, drafted in accordance with an attached model form specifying the final destination, the quantities, the approximate date of delivery, the timetable for utilisation, the form in which delivery is to take place and the allocation of the material to one or other of the uses specified in the Declaration: fuel supply for a reactor, fabrication of fuel elements, research and development, or utilisation in any other installation connected with an energy or research and development programme; these requirements also apply to any intermediate storage;
- the government of the Member State to which the consignee belongs must confirm the accuracy of the information contained in the said certificate;
- such plutonium and uranium will not be retransferred to a third State without mutual agreement between the Member State that has separated or enriched them and the Member State desiring to receive the retransferred material, "without prejudice to any other rights of prior consent that may exist."

---

[60] Nuclear Energy Agency, 'The Regulation of Nuclear Trade: Non-Proliferation—Supply-Safety: International Aspects,' Volume 2, 1988, p. 84.

Exceptions have, however, been made in the case of plutonium with an isotopic concentration of plutonium 238 above 80 % and of special fissile materials used in quantities of the order of 1 gram as a component of measuring instruments.[61]

5.2. *Notifications to the IAEA of Exports and Imports of Nuclear Material by NWS*

With a view to helping the Agency with its safeguards activities, the Resident Representatives of the US, the UK and the Soviet Union informed the DG of the IAEA, on 10th July 1974, of the decision of their respective governments henceforth to communicate to the Agency information concerning their exports and imports of nuclear materials (INFCIRC/207–26th July 1974). By the letter of 7th February 1984, the Governor representing France on the BG informed the DG that the French Government would do the same (INFCIRC/207/Add. 1-May 1984).[62]

5.3. *Final Declaration of the Conference of Ministers of Foreign Affairs of the Non-Aligned Countries, September 1985*

The conditions imposed on nuclear trade by the exporting countries of the Zangger Committee and the London Club gave rise to strong criticism from developing countries, and especially from "newly industrialised countries" with ambitious nuclear programmes, such as India, Brazil and Argentina. These criticisms were aired in the form of certain principles debated in the IAEA Committee on Assurances of Supply (CAS) and on the occasion of NPT Review Conferences.[63]

Similar criticisms were also voiced at the Conference of Ministers of Foreign Affairs of the Non-Aligned Countries held at Luanda from 4 to 7 September 1985. Expressing their "concern at the obstacles which the developed countries place in the way of transfer of technology related to the peaceful uses of nuclear energy by fixing conditions which are incompatible with the sovereignty of developing countries," the Ministers "affirmed the inalienable right of all States to apply and develop their programmes for peaceful uses of nuclear energy for economic and social development, in conformity with their priorities, interests and needs."

[61] Ibid.
[62] Ibid, pp. 84–85.
[63] Ibid, p. 85.

They added that it was necessary to comply with the principles of non-discrimination and free access to nuclear technology; non-proliferation should not be used as an excuse to prevent States from exercising their rights fully (INFCIRC/332, March 1986).[64]

6. *Role of the Emerging Nuclear Suppliers and Non-Proliferation*

6.1. *Introduction*

The emerging suppliers' nuclear programs were developed because of a mixture of perceived development needs, prestige, a desire for energy self-reliance, and potential security options. Many of these programs were initiated with the full support and encouragement of the US and other traditional nuclear suppliers. In the mid-1970s, when concerns over proliferation increased, access to nuclear technology became a much more sensitive issue. Controls were placed on certain exports and end uses. There was implicit denial of sensitive nuclear materials to certain states, and the IAEA safeguards and monitoring system was expanded and refined.[65]

However, the number of states capable of exporting nuclear material, technology, equipment, and services is growing. Once confined primarily to States party to the NPT, the list of actual and potential nuclear suppliers now includes many countries that do not subscribe to the NPT or to other international nuclear export control agreements. Although international control accords—such as the Zangger Committee and the NSG guidelines—do not prohibit the export of sensitive nuclear materials and equipment, they do reduce the risks of proliferation by imposing international safeguards as a condition for export. It is the concern of many non-proliferation experts that the emergence of new nuclear suppliers that are not bound by existing international controls could erode the system of export restraints and threaten the viability of the non-proliferation regime.[66]

---

[64] Ibid.

[65] Sands, Amy, 'Emerging Nuclear Suppliers: What's The Beef?' in William C. Potter (ed.), 'International Nuclear Trade and Non-Proliferation—The Challenge of The Emerging Suppliers,' 1990, p. 31.

[66] Potter, William C., 'International Trade and Non-Proliferation: The Challenge of the Emerging Suppliers,' 1990, p. 3.

To date, concern regarding the issue of the emerging nuclear suppliers and non-proliferation has not been matched by systematic research on the topic. Indeed, there is little agreement on who the emerging suppliers are, why they have entered the international nuclear market, and the extent to which they have departed from the export practices of the more established or so-called traditional Supplier States.[67]

Despite efforts to prevent nuclear proliferation, the interest in developing civilian nuclear energy programs continued to be encouraged and assisted. The NWS and advanced industrial states seemed to want to hook developing countries with the promise of nuclear energy's development benefits without encouraging them to develop the indigenous technical expertise. In other words, developing countries were expected to remain dependent on the original suppliers.[68]

As might be expected, several developing countries saw this as an issue of sovereignty and long ago decided to develop their own nuclear capabilities. Relying on the non-proliferation regime's claim that it was not nuclear energy but weapons development that was being limited, such countries as India, South Korea, and Argentina have taken full advantage of the nuclear cooperation, training, and technology transfer allowed via IAEA programs and permitted by the NPT. Several countries intensified their own efforts to develop a complete fuel cycle and nuclear self-sufficiency after the US unilaterally tightened its nuclear export controls to require full-scope safeguards on its nuclear exports. Countries did not want to be vulnerable to American policy shifts.[69]

During the 80's, several Third World countries called for developing countries to cooperate in the field of nuclear research. As a consequence, Pakistan signed a nuclear agreement with Malaysia; Argentina signed a contract for the sale of enriched uranium to Iran and later to Algeria; and in 1988 it signed a nuclear cooperation agreement covering research and development in the nuclear fuel cycle with Egypt.[70]

Lacking experience and proven technical capabilities, emerging suppliers promote their nuclear exports with a variety of tactics. They can

[67] Ibid.

[68] Sands, Amy, 'Emerging Nuclear Suppliers: What's The Beef?' in William C. Potter (ed.), 'International Nuclear Trade and Non-Proliferation—The Challenge of The Emerging Suppliers,' 1990, p. 31.

[69] Ibid.

[70] Rydell, Randy J., 'Studying the Emerging Nuclear Suppliers: A Framework for Analysis,' in William C. Potter (ed.), 'International Nuclear Trade and Non-Proliferation—The Challenge of The Emerging Suppliers,' 1990, p. 13.

under-price their products to get a foothold in the market. They also might undertake transactions in politically uncertain areas to show support for their "brothers," to prevent the domination of traditional suppliers, or for purely commercial reasons. Argentina has already built a pilot fuel fabrication plant in Chile and has contracted to build a research reactor in Algeria. South Africa has long been an exporter of uranium to Western European countries, Japan, and Taiwan. What would prevent Pakistan from helping Saudi Arabia or even Libya at some point in the future?[71]

In theory, there is nothing that prevents Pakistan from helping either Saudi Arabia or Libya to develop nuclear capabilities; however, in practice, the technological gaps between these countries are massive, which means, it will not be feasible for them to acquire such technology—from Pakistan or anywhere else—in the near future. Although, the latest news in early 2000s about Khan's network to sell nuclear technology to Libya proved that this could take place. Nevertheless, the technological gap between Pakistan and Libya in the nuclear field leaves the latter with material and equipment with no real potential for development.

Emerging suppliers may not always find it in their best interest, however, to provide sensitive nuclear items to potential customers. First, emerging suppliers remain dependent on foreigners for various types of technology, materials, and assistance. Even China, a recognised NWS, has acknowledged its need for advanced nuclear technology for its civilian energy program and has tempered its international nuclear transactions to fall more within the non-proliferation regime's umbrella. Going too far in providing sensitive nuclear materials or technology to potential proliferants might result in the cut-off of an emerging supplier from a much-needed foreign technology.[72]

Second, emerging nuclear suppliers may not see any political advantage from selling its nuclear technology if it is perceived as a national asset not to be passed on lightly. The supplier's national security and political interests may play a greater role in limiting to whom and under what circumstances nuclear technology and materials are exported. Argentina might require more control on nuclear items exported to Chile than to Iran, for example, or it might decide against exporting anything. Just as

---

[71] Sands, Amy, 'Emerging Nuclear Suppliers: What's The Beef?' in William C. Potter (ed.), 'International Nuclear Trade and Non-Proliferation—The Challenge of The Emerging Suppliers,' 1990, p. 32.

[72] Ibid.

traditional suppliers balance political and commercial factors in deciding what business to pursue, emerging suppliers may also be sensitive to trade-offs that promote security over commercial gains.[73]

Finally, having newly acquired technical capabilities and limited financial resources, emerging nuclear suppliers might not fully understand or know their own limits. With too strong an export push, a country could overextend its own resources. Most emerging suppliers do not have much depth in skilled personnel, particularly in the middle management and engineering areas, nor do they have extensive funds. Any delays or complications would increase costs and use resources needed elsewhere. Commitment to foreign projects that drain these already scarce resources could hurt a country's domestic nuclear industry.[74]

6.2. *Emerging Nuclear Suppliers*

Suppliers differ significantly in their technical capabilities, experience, and level of regime involvement, and these distinctions result in different nuclear export policies.[75]

Several nuclear suppliers that are relatively new to the international nuclear marketplace could challenge the market share of the traditional suppliers because they have advanced nuclear capabilities. Countries such as Japan, China, and Spain fall into this category, as might South Korea and Taiwan, whose capabilities are sophisticated but limited. These states may not be fully incorporated into the supplier regime, but they now claim to accept many of the rules of the regime. Nevertheless, because of the past activities and policies of some of these new but advanced nuclear suppliers, their current activities remain a concern. The capabilities, policies, and behaviour of this group of emerging suppliers need to be monitored, examined, and analysed.[76]

Another concern involves a possible *new* generation of nuclear suppliers and the likelihood that these "emerging suppliers" will not easily, if at all, be lured into the current non-proliferation regime. For the most part, these emerging suppliers are the countries that have been developing indigenous, unsafeguarded capabilities and are often thought of as potential proliferants. Several States, such as India and Pakistan, have

[73] Ibid.
[74] Ibid, pp. 32–33.
[75] Ibid, p. 25.
[76] Ibid, p. 27.

declared that the regime is discriminatory and is an effort by the weapons states to maintain their dominance. Thus, some of these new suppliers reject the current non-proliferation regime and sometimes act outside its control system; they also have either limited nuclear capabilities or rudimentary ones.[77] North Korea and Israel can also be added to the list of emerging suppliers who reject the non-proliferation regime.

In fact, India is involved in several bilateral programs regarding research and development with third parties like Israel. Meanwhile, Pakistan has been cooperating with North Korea to develop their nuclear capabilities and the means of delivering nuclear weapons. According to US officials, Pakistan was a key supplier to North Korea's secret nuclear programme.[78] Mansoor Ijaz and R. James Woolsey stated that, "Islamabad vehemently denies having provided North Korea with any nuclear assistance in the past, but mounting intelligence data and forensic evidence suggest otherwise. The same South Korean intelligence report that exposed the existence of the uranium enrichment facility that sparked the US confrontation with North Korea last fall also reportedly noted remarkable similarities between centrifuge components bought by North Korea for its plant and those known to be used by Pakistan at its enrichment facilities."[79] The American State Department stated, "Many have studied nuclear technology (though not necessarily nuclear weapons production) in the Soviet Union and China and reportedly Pakistan."[80] So it is not fully appropriate to say that they have limited nuclear capabilities; those countries might have limited nuclear capabilities compared to the US or France, but they are quite advanced compared to developing countries. However, it should be a priority to give attention to the small suppliers as well as the big suppliers because they are all proliferants in this context.

Emerging nuclear suppliers—defined as including both the new (but advanced) suppliers and the possible new generation of suppliers—do represent a threat to the non-proliferation regime; however, the reality of the threat from their potential exports has sometimes been exaggerated. In seeking business opportunities, potential new suppliers market the

---

[77] Ibid.

[78] Goldenberg, Suzanne, 'Pakistan Helped North Korea Make Bomb,' 19 October 2002.

[79] Ijaz, Mansoor, and R. James Woolsey, 'Cut Supply Lines that Fuel Pyongyang's Nuclear Dreams,' 12 January 2003.

[80] Niksch, Larry A., 'North Korea's Nuclear Weapons Program,' State Department, CRS Issue Brief for Congress, updated 10 March 2002 16 May 2002 p. 11.

most optimistic version of their capabilities. Obtaining a realistic assessment of their actual capabilities—and thus of the threat these new players present to the non-proliferation regime—requires examining a country's real technical capabilities versus its declared technical capabilities as well as its international obligations and domestic regulations, its domestic economic and political circumstances, and its current and likely future activities.[81]

Role of the Emerging Suppliers in the Nuclear Market

In attempting to evaluate the threat of the emerging nuclear suppliers, it is crucial to define the scope of activities and the range of materials, equipment, and technology that are of concern. Nuclear trade of any kind could be considered an issue because of its potential to evolve into more sensitive areas or because of the links it encourages between countries outside the non-proliferation regime. However, some attempts at ranking the significance of different transactions is possible and helpful. Clearly, trade in nuclear power plants is not as significant as trade in weapons-grade materials or technology that might be used to fabricate nuclear weapons.[82]

The nuclear marketplace has several different levels, each possibly playing a role in the development of nuclear weapons, but with certain aspects being more critical than others. At the moment, transactions within the nuclear marketplace tend to concentrate on services, small-scale projects, research and development (R&D) activities, the use of radioactive materials in medicine and agriculture, as well as training. Although limited commerce in large-scale, costly nuclear power plants and fuel cycle facilities is occurring, some do take place (for example, in China, South Korea, and possibly Indonesia), and extensive effort is being put into attempting to obtain the few large projects under discussion. Possible categories for these transactions are as follows: nuclear power plant, fuel cycle facility, R&D, materials services, non-energy-related use of nuclear, materials/technology, and optional nuclear weapons use.[83]

---

[81] Sands, Amy, 'Emerging Nuclear Suppliers: What's The Beef?' in William C. Potter (ed.), 'International Nuclear Trade and Non-Proliferation—The Challenge of The Emerging Suppliers,' 1990, p. 26.

[82] Ibid, pp. 27–28.

[83] Ibid, p. 28.

The reality of today's marketplace is that most potential proliferants lack financial resources to pursue large-scale projects and most suppliers cannot organise alternative funding methods. The emphasis is on small projects that are affordable or that fall within some international agency's arena of interests. It would be a mistake, however, to underestimate the importance of this "slow period" in nuclear transactions; Countries such as Pakistan did not need large-scale projects to overcome the remaining barriers in their nuclear weapons effort. Specific components, certain materials (for example, maraging steel), or training may provide the missing skills or items.[84] Pakistan was able to overcome those barriers through bilateral cooperation with nearby States like China and North Korea. The American State Department mentioned that "In February 1996, the US press reported leaked US intelligence reports alleging that China sold ring magnets to Pakistan in 1995 that could be used in enriching uranium for nuclear weapons."[85] This is an indication of the help Pakistan has received from neighbouring States.

Emerging suppliers provide certain advantages to a buyer. They offer attractive and less restrictive terms. Services, technology, equipment, or materials can be supplied with minimal safeguards requirements, fewer restrictions on end use, and less cumbersome export licensing procedures. The price for items could also be less than what a traditional supplier might offer. They might also be willing to do business in politically sensitive areas. A recent example of this possibility was Iran, where Argentina's ENACE—jointly owned by the Argentine National Atomic Energy Commission and the German firm Kraftwerk Union (KWU)—negotiated to finish the construction of the Iranian nuclear power plant at Bushehr. Most countries and individual companies refused to do nuclear-related business with Iran until the Iran-Iraq war ended. However, Argentina, acting as the lead negotiator for a consortium of companies, appeared willing to take the flak (political and real) to get the deal. In this case, Argentina would have provided access for KWU to a market possibly denied by German policies. Iran, for its part, appeared willing to rely on Argentina, because it did not have many other options recognising the technical strength of the other members of the consortium.[86] Iran contacted Argentina in 1984 to acquire highly enriched uranium for its

---

[84] Ibid, p. 29.

[85] Blood, Peter R., 'Pakistan-US Relations,' State Department CRS Issue Brief for Congress, updated 10 March 2002, p. 4.

[86] Ibid, pp. 29–30.

research reactor in Tehran and as a potential supplier to its power plant in Bushehr.[87] However, Iran contacted Russia at a later stage to finish building the Bushehr power plant.

Another concern for buyers is that their credibility as States, interested in peaceful nuclear development, could be jeopardised by using emerging suppliers that have fewer export constraints. Regardless of the official justification for buying from Argentina or India, such a purchase could cause other countries to be suspicious of the motives of the buyer and to question nuclear exports to that State in the future.[88]

However, recent events have challenged this view, suggesting that which supplier—major or emerging—an NNWS approaches to develop its nuclear infrastructure does not matter: the US is still likely to express suspicion regarding the intentions of an NNWS which seeks to develop its nuclear capabilities for peaceful purposes. This became evident when the US raised suspicions around the nuclear programme of Iran. Iran had approached Russia, a major supplier, to build its nuclear power plant at Bushehr, but this did not prevent a US- led campaign to stop Iran from developing its nuclear capabilities for peaceful purposes.

Regardless of the scientific ability or technical experience of the Supplier State, potential buyers from developing States may see one of these emerging suppliers as more appealing than the traditional ones. Buying a reactor from Argentina may be more cost-effective than buying it from traditional suppliers like the US or Canada. So even if a traditional and an emerging supplier are both willing to sell a reactor, the developing NNWS may choose to strike a deal with an emerging supplier, being more affordable.

Threat of Proliferation Caused by the Emerging Suppliers

The current problems of the non-proliferation regime reflect the tensions resulting from the historic prestige attached to nuclear weapons and technology, the perceived security benefits of having nuclear weapons, the parallel development of nuclear weapons and energy programs and

[87] Timmerman, Kenneth R., 'Iran's Nuclear Program: Myth and Reality,' 1995, a paper presented to The Italian Union of Scientists for Disarmament (USPID) Sixth International Castiglioncello Conference entitled 'Fifty Years after Hiroshima,' 30 September 1995.

[88] Sands, Amy, 'Emerging Nuclear Suppliers: What's The Beef?' in William C. Potter (ed.), 'International Nuclear Trade and Non-Proliferation—The Challenge of The Emerging Suppliers,' 1990, p. 33.

the increasing concern among traditional suppliers about the diffusion of sensitive nuclear technology. From the developing States' perspective, it appeared to be acceptable for Japan, Italy, and even Spain to become reasonably self-sufficient nuclear States, but not for Brazil, South Korea, or Iraq. In a way, "the political realities of the regime may themselves provide an incentive for proliferation," especially now that the effort to limit proliferation appears to include an effort to prevent States from obtaining a "nuclear weapons option" as well as a nuclear weapon. Thus, States that show interest in developing a complete fuel cycle become suspects, especially if they have not signed the NPT.[89] In addition, even if a State is a member of the NPT and its nuclear facilities are subject to the IAEA safeguards, once it decides to develop its nuclear capabilities for peaceful purposes, it may be accused of trying to build nuclear weapons, as in the case of Iran.

The rules of the NPT and the Guidelines of the NSG do not ban the transfer of nuclear technology, but they do restrict it. Since most of the new suppliers are members of the NSG, and they apply the Guidelines of the Group in their nuclear transactions, they are not proliferants any more than the traditional suppliers are.

Regarding the implementation of the Guidelines, developing States view them as creating double standards and their implementation confirms this view. The nuclear suppliers have prepared a black list of States that should not receive any nuclear assistance from members of the Group and such a list is based on political interests not rules.

The challenge of emerging suppliers may be the catalyst either for the reform of the non-proliferation regime or its destruction. As long as emerging nuclear suppliers are outcasts from the supplier regime, they present a serious destabilising force to non-proliferation efforts. Their ability and willingness to circumvent NSG guidelines and to offer items with fewer restrictions could undermine the current shaky consensus. Their being outcasts permits them to question the legitimacy of the regime and to act against it. Before their capabilities catch up to their promises and they become credible alternative sources of nuclear technology and materials, the supplier regime must be adapted and expanded to conform to the realities of today's complex international nuclear marketplace.[90]

---

[89] Ibid, p. 35.
[90] Ibid, p. 36.

7. *Future of the Supply Regime*

The differences in interests, interpretation, and implementation among the Supplier States, along with commercial competition and inadequately regulated industry actions, have been exploited by States seeking nuclear weapons or seeking the capabilities to produce them. One recent example is noteworthy. German and Swiss authorities had been investigating alleged export violations involving the theft of Urenco enrichment technology and the attempted transfer of that technology and equipment to Pakistan. Leybold officials in the FRG and Switzerland were alleged to have aided the production of enrichment equipment destined for export to Pakistan. In January 1986, Swiss customs authorities confiscated three autoclaves and other equipment especially designed by the Swiss firm Mettalwerke Buchs AG and other centrifuge equipment, along with the blueprints used to produce them. These were allegedly stolen from URANIT GmbH, the West German partner in Urenco, by officials from Leybold-Hereaeus. In May 1987, following a raid by West German customs authorities, the Aachen State prosecutor's office announced that it was conducting a criminal investigation of Leybold-Hereaeus, suspected of illegally providing assistance to the uranium enrichment activities of Pakistan. Hence, looking at the future of the supply regime and the supplier institutions in which it is embodied, there are a number of disturbing trends that could exacerbate such supplier problems.[91]

The emerging suppliers have explicitly challenged the regime, arguing that it is ineffective, violated by the NWS and the nuclear suppliers, and is discriminatory. Many of them are hostile to the regime partly because they inevitably perceive the proliferation danger from a different perspective from either the US or other major suppliers. They feel that in the long term such modest measures will not be sufficient, and they argue for major changes that directly or indirectly challenge the scope and structure of what they view as a discriminatory regime.[92]

More direct nuclear supply issues are affected by controversies over the benefits and opportunities offered by the peaceful application of nuclear energy disputing over which activities are peaceful. In this context, there is disagreement (political, not technical) over sensitive nuclear technologies, with disputes over the question of whether to make enriched

[91] Potter, William C., 'International Trade and Non-Proliferation: The Challenge of the Emerging Suppliers,' 1990, p. 57.

[92] Ibid, pp. 57–62.

uranium and separated plutonium available to States that do or do not have an apparent need for these materials for advanced nuclear research or power programs. There are those who argue that dispersal of these products exacerbates the dangers of nuclear proliferation and terrorism and that their spread should be severely restricted; they argue that if sensitive nuclear materials are to be made available, this should be done only under appropriate safeguards, and they assert that only such measures will keep other countries from developing sensitive nuclear technologies themselves. But there are those who argue for an intrinsic right to such technologies, asserting that any controls other than nominal safeguards are unnecessary and unnecessarily limiting and that technology transfers should facilitate rather than preclude indigenous development efforts. It would appear that these uncertainties, complexities, and ambiguities could be exacerbated if the emerging suppliers enter the market in a significant way, even if they insist on safeguards for their exports, but do not back away from their challenge to the commitments of the NPT, Tlatelolco, or other elements of the non-proliferation regime. If this occurs, efforts to prevent proliferation could become more ineffective and could undermine the establishment of a broad consensus on rules of the game for nuclear supply. However, it is clear that such consensus is necessary if the supply and non-proliferation regimes are to be effective.[93] This consensus can be reached through broadening the international participation in the regime.

The experience of the Peaceful Uses of Nuclear Energy (PUNE) Conference, where such attacks were heard again and again, is illustrative. A scathing attack on the NPT was launched by Iran, an NPT party, which asserted that the treaty was ineffective and subject to numerous violations. In a PUNE document, Iran posed the following charges:[94]

1. The NPT has not prevented the vertical proliferation of nuclear weapons.
2. The NPT has not prevented the spread of nuclear weapons, namely the deployment of nuclear arsenals in NNWS (spatial proliferation).
3. The NPT has not facilitated international cooperation in the peaceful uses of nuclear energy.
4. The objectives of NPT specially Articles I, II, IV and VI have been violated by many States party to the treaty.

---

[93] Ibid, p. 63.
[94] Ibid, pp. 66–67.

5. The NPT has not been able to facilitate the establishment of genuine NWFZ in the world.
6. The claim that no more NWS has been added to the existing five States, since NPT entered into force, is to divert attention from the fact that there are States such as occupying regime of Palestine, so called Israel, and the racist regime of South Africa which have been granted full access to nuclear weapon technology and materials by some of the nuclear weapon States, especially the US.
7. The NPT has not been able to prevent the militarisation of International Waters made by nuclear submarines.

8. *Conclusions*

The supplier groups are regarded by their founders, i.e., the Supplier States, as a complementary means to prevent nuclear weapons proliferation. However, NNWS, especially the developing countries, regard the supplier groups as means of control intended to maintain the monopoly of nuclear supplies in the hands of the suppliers.

There is a major question in this regard: has the work of these groups affected negatively the transfer of nuclear technology for peaceful purposes? The answer is "YES."

Although there are no statistics in this regard, the mere fact that the nuclear suppliers' requirements exceed those of the NPT and the IAEA as such comprises a negative effect on supplies.

Some specialists in the field have confirmed this position, stating that the first reply to any request of supply is "NO," at least until the NNWS proves its good will and that it fulfils the criterion of non-diversion of the material and equipment into military uses as required by the Supplier State. Mr. Jacques Baute, Director of the Iraq Nuclear Verification Office at the IAEA, stated in a meeting in Vienna in May 2004, "Although there are no direct relations between the Agency and the Supplier Groups, the latter are known to deny the supply of certain technologies. There are many denials, which in turn leads to the frustration of the party which has been denied the technology. The denials are based on proliferation threats or political reasons. The real problem lies in which country asks for supply, so there is a need for a multilateral treaty to control the supply—not a club of several States." He added, "While studying the case of Iraq, we have found out that the trigger lists were reasonable. The included items are all dual-use items. We went through Annex 3 to ensure that these suppliers would inform the Agency of any supplies to Iraq. Some of these items are interesting from the perspective of proliferation. Hence, they increase the proliferation risks. However, at a later stage politics interferes in choosing which States are getting supplies."[95]

In an interview conducted in Vienna, Mr. Tarik Raouf, Head of Verification and Security Policy Cooperation at the IAEA, confirmed this position when he stated, "There are no statistics available on denials of supply requests by the NSG, these are confidential issues for them. Their rules absolutely affected negatively the transfer of nuclear technology around the world. If the requesting State is not a member of the NSG, then the presumption is denial. According to the information the NSG provided to experts in a meeting in 1997, they mentioned a percentage of supply and denials. It was a relatively small number of requests that has been denied."[96]

Nevertheless, NSG sometimes attempts to impose embargoes. The Group does not make formal decisions to boycott countries, but any member State whose policy calls for an embargo would use the NSG forum to encourage others to do the same.[97]

The main goal of the supplier groups is to restrict the flow of nuclear materials and equipment around the world. Hence, a State requesting supplies has to prove not only that it is a member of the NPT and that its nuclear facilities fall under IAEA safeguards, but also that the practice of the NPT/IAEA safeguards on its territory is sufficient to prevent any misuse of the nuclear material and equipment transferred from the point of view of the supplier State. This is a subjective criterion that the Supplier States members in the nuclear suppliers' groups decided to implement while considering any request of supply.

The previous stand constitutes a violation by the suppliers of their legal obligations under the NPT. All NPT obligations are objective and cannot be qualified, modified or curtailed by subjective interpretation.[98]

---

[95] Baute, Jacques, Director of Iraq Nuclear Verification Office, in an interview held at the IAEA Headquarters in Vienna, May 2004.

[96] Statement of Raouf, Tarik, Section Head, Verification and Security Policy Coordination, Office of External Relations and Policy Coordination at the IAEA, in an interview concluded at IAEA Headquarters, 2004.

[97] Nasseri, Cyrus, 'A Perception for Evolution: the NSG's Impact on Non-Proliferation and the Right to Access,' 1999, p. 51.

[98] Ibid, p. 56.

CHAPTER FOUR

CASE I: IRAQ

The Iraqi nuclear programme was initiated as early as the 1950s. The programme commenced peacefully, but when the intentions of the political leadership changed, its aim shifted to exploring the military capability of nuclear technology. From the 1970s onwards, Iraq, as a wealthy oil producing country, continuously developed its capabilities in all fields—including nuclear technology.

Iraq signed the NPT on 1 July 1968 and ratified it on 29 October 1969. It then signed a Safeguards Agreement with the IAEA (INFCIRC/153) in 1972. Its nuclear facilities were inspected regularly by the IAEA to ensure the compliance of Iraq with the NPT, i.e., that it was not employing its nuclear facilities for military purposes. However, Iraq discreetly managed to start a clandestine nuclear programme about which the IAEA knew nothing for many years.

In this chapter, the following shall be discussed:

- I. Development of the Iraqi nuclear programme.
- II. Role of World Nuclear Suppliers in developing the Iraqi nuclear programme.
- III. Israeli bombing of Osirak.
- IV. Iraqi violations of its international obligations.
- V. Process of disarming Iraq under UN SC resolutions.

1. *Development of the Iraqi Nuclear Programme*

The introduction of Iraq to the field of nuclear technology began during the fifties with the aim of harnessing the atom to reap the benefits of cheap energy and to enhance medical research. The Iraqi Government, together with allies in Turkey and Pakistan, funded a research centre in Baghdad which was in actuality no more than a spartan collection of laboratories. In 1958, a new government came into power in Iraq. It proceeds to strike a deal with Moscow for the establishment of forty industrial projects, the centrepiece being a small two-megawatt research reactor, the IRT 2000, built at Tuwaitha, sixteen kilometres south-east of Baghdad.[1]

During the late 70s, Iraq supported the training of its scientists around the world, especially in the UK, France and Italy.[2] Iraq's secret nuclear programme (known as "Petrochemical Three" (PC3)) was conducted under the authority of its Atomic Energy Commission with support from the Defence Ministry and the Ministry of Industry and Military Industrialisation. PC3 received massive infusions of money and resources in order to build a nuclear explosive device.[3]

The immediate objective was to acquire nuclear technology, but the long-range objective of the plan was to produce nuclear weapons. Iraq had impeccable credentials for receiving nuclear assistance. As an NPT party hosting IAEA inspections of its declared facilities on a regular basis, Iraq raised little doubts about its nuclear projects.[4]

Iraq manoeuvred its way through the IAEA system. In 1974, it assigned a "scientific attaché" to its Embassy in Vienna whose main job was to provide detailed reports on IAEA matters, including the role of inspectors in uncovering clandestine programmes and the control over information within the organisation. Furthermore, Iraqi inspectors at the IAEA informed their government of the latest tools used during inspections.[5]

Hence, the Iraqi regime acquired sufficient inside knowledge about the IAEA system, and was able to find means to cover up the clandestine activities within its nuclear programme. During the same period, Iraqi scientists started to shop for nuclear material and equipment to expand nuclear projects in order to produce enough fuel to build a nuclear explosive device.

In 1975, the Iraqi and French governments concluded a Treaty on Nuclear Cooperation that permitted the purchase of two French reactors fuelled by HEU (Osirak). Hot cells that would permit irradiating natural uranium to produce some plutonium were installed. A further agreement

---

[1] Bhatia, Shyam and Daniel McGrory, 'Brighter than the Baghdad Sun- Saddam Hussein's Nuclear Threat to the US,' 2000, p. 56.

[2] Hamza, Khidhir, 'Inside Saddam's Secret Nuclear Program,' Bulletin of the Atomic Scientists, Volume 54, No. 5, September/October 1998.

[3] Albright, David and Mark Hibbs, 'Iraq's Shop-Till-You-Drop Nuclear Program,' Bulletin of Atomic Scientist, Volume 42, no. 3, April 1992.

[4] Deaver, Micheal V., 'Disarming Iraq, Monitoring Power and Resistance,' 2001 p. 35.

[5] Hamza, Khidir with Jeff Stein, 'Saddam's Bombmaker, The Terrifying Inside Story of the Iraqi Nuclear and Biological Weapons Agenda,' 2000, pp. 98–100.

with Italy provided Iraq with a fuel reprocessing plant in which plutonium could be extracted. The Israeli air strike on Osirak in 1981 brought direct nuclear cooperation with France and Italy to an end.[6]

Iraq bought hot cells that could handle uranium from Italy. Control panels were purchased from Brazil, valves from France and copper cooling coils from Finland. It ordered uranium from Niger, Italy[7] and Portugal and it hired construction firms from Yugoslavia. Two Swiss firms sent components for centrifuges, including parts manufactured from 'maraging steel' that were strong enough to withstand crushing vibration. German companies were heavily involved in supplying Iraq with nuclear materials and equipment especially in developing its enrichment of uranium efforts, among which were H&H, ROSCH and MAN.[8] Iraq also bought a nuclear accelerator from the National Electrostatic Corporation in Madison, Wisconsin.[9]

Factories, electrical supply and power equipment which were some of the basic infrastructure Iraq needed for the programme were easy to obtain from abroad. In addition, to evade export controls, Iraqi officials divided equipment orders into innocuous sub-components. They also tried to buy machines to produce sensitive components in Iraq. They used intermediaries to disguise the destination of equipment and materials. The Iraqis seized opportunities to obtain any uncontrolled equipment and material that was available, often in enormous quantities, regardless of whether they were ready to use or not. Their strategy was to buy whenever there was an opportunity and simply run the risk that some material might not be used. That practice generated the impression later on that Baghdad was closer to having a nuclear bomb than it in fact was.[10]

---

[6] NPT Review Conference, NPT/CONF.2000/28, 'Final Document of the 2000 Review Conference of the Parties to the NPT,' 24 May 2000, p. 40.

[7] SC, S/23644, 'Note by the UN SG,' annexed to it a letter dated 25 February 1992 from the DG of the IAEA addressed to the SG attaching 'the Report on the Tenth IAEA on-Site Inspection in Iraq (5–13 February 1992) under SC Resolution 687 (1991) SC,' 26 February 1992.

[8] SC, S/1997/779, attached to it the 'Letter dated 6 October 1997 from the DG of the IAEA' annexing to it the 'Fourth Consolidated Report of the DG of the IAEA Under Paragraph 16 of the SC Resolution 1051 (1996),' 8 October 1997, p. 39.

[9] Hamza, Khidir with Jeff Stein, 'Saddam's Bombmaker, The Terrifying Inside Story of the Iraqi Nuclear and Biological Weapons Agenda,' 2000, p. 98.

[10] Albright, David and Mark Hibbs, 'Iraq's Shop-Till-You-Drop Nuclear Program,' Bulletin of Atomic Scientist, Volume 42, no. 3, April 1992.

The Iraqi Government abandoned the plutonium route in favour of HEU. Because the fuel in the Iraqi reactors was safeguarded, a separate supply of natural uranium had to be enriched in order to create the needed nuclear material for a device. Two sources were found for natural uranium: it was extracted from Iraqi phosphates and hundreds of tons were imported from Brazil,[11] Portugal, and Niger.[12] Faced with tighter export controls, they proceeded with large procurements as opportunities presented themselves, even though they had no immediate plans for materials for the quantities ordered.[13]

The revelation of the Iraqi military programmes prompted companies that contributed to Saddam's effort to deny any affiliation to the Iraqi clandestine nuclear programme. But the specific nature of some of the equipment exported should have aroused suspicions of companies and export control authorities. According to the eighth IAEA report, in some cases "the presence of application-specific fixtures removes most doubt as to the intended use." The intermediaries or companies that dealt with Iraq "must have known (or could reasonably have inferred) the intended uses."[14]

Iraq declared it had acquired 700 uranium hexafluoride-resistant bellows valves from Swiss companies Balzer AG and VAT AG, and the US Company Nupro. It also purchased (but refused to reveal the supplier of) 100 tons of maraging steel, sold as both pre-formed cylinders ("performs") and in sheets. The material is believed to have come from Germany or Britain.[15]

Until 1995, Iraq denied having had any serious intention of building nuclear weapons. Then, after the defection of Hussein Kamel, Saddam's son-in-law and head of the Ministry of Industry and Military Industrial-

---

[11] SC, S/22986, 'Note by the UN SG to the SC,' Annexed to it Letter dated 27 August 1991 from the DG of the IAEA addressed to the SG attached to it the 'Report on the Fourth IAEA on Site Inspection In Iraq (27 July–10 August 1991) under Security Council Resolution 687 (1991) under SC Resolution 687 (1991),' SC, S/22986, 28 August 1991.

[12] Evans, Gareth, 'Statement to the 1995 Review and Extension Conference of the States Parties to the Treaty on Non-Proliferation of Nuclear Weapons,' 1995, p. 26.

[13] SC, S/23505, 'Note by the SG to the SC,' Annexed to it Letter dated 28 January 1992 from the DG of the IAEA addressed to the SG attached to it the 'Report on the Ninth IAEA on-Site Inspection in Iraq (11–14 January 1992) under SC Resolution 687 (1991),' 30 January 1992, p. 9.

[14] SC, S/23283, 'Note by the UN SG,' transmitting the letter dated from the DG of the IAEA addressed to the SG annexed to it the 'Report on the Eighth IAEA on Site Inspection in Iraq (11–18 November 1991) under the SC Resolution 687 (1991),' 11 December 1991, p. 14.

[15] Albright, David and Mark Hibbs, 'Iraq's Shop-Till-You-Drop Nuclear Program,' Bulletin of Atomic Scientist, Volume 42, no. 3, April 1992.

isation, in August 1995, his revelations assisted the IAEA to discover the scope and intensity of the Iraqi nuclear weapons programme.[16]

The inspectors eventually found out that the principle nuclear research centre in Iraq was Al-Tuwaitha. Theoretical studies, based on mainframes and personal computers, electrical design studies, development dedicated instrumentation, radio-chemistry experiments, separation of a few grams of plutonium, studies of uranium metal production and casting were conducted as part of the activities related to fuel fabrication at the site. There were buildings constructed to house R&D of both EMIS and gaseous diffusion enrichment programmes.[17]

Al Atheer Centre was designed to accommodate all technical activities related to nuclear weapon development, including experiments with high explosives, uranium metallurgy studies and fabrication, the machining of uranium metal, device assembly and storage.[18]

Al Qa Qaa facility had the infrastructure to support initial activities of the development of a high explosive package for a nuclear weapon.[19]

Al Tarmiah and Al Sharqat complexes were designed for housing the Calutron separators for EMIS.[20]

What Iraq lacked was a complete nuclear core. It had more than twenty-five kilograms of bomb-grade uranium fuel rescued from Osirak, the French reactor bombed by Israeli jets in 1981. Twelve kilograms were 93% -enriched uranium, and about fourteen kilograms were enriched to 80%. Additional uranium was available from the irradiated fuel of the Russian reactor (although it would have been too hot to handle).[21] Altogether, this would have been more than sufficient to produce the eighteen to twenty kilograms needed for a bomb. The breaking of Desert Storm, however, did not leave sufficient time for uranium metal to be extracted from the fuel. Instead, Saddam ordered his Special Security Organisation to take possession of the bomb components and conceal them from foreign inspectors.[22]

---

[16] Hamza, Khidhir, 'Inside Saddam's Secret Nuclear Program,' Bulletin of the Atomic Scientist, Volume 54, No. 5, September/October 1998.

[17] SC, S/1997/779, attached to it the 'Letter dated 6 October 1997 from the DG of the IAEA' annexing to it the 'Fourth Consolidated Report of the DG of the IAEA Under Paragraph 16 of the SC Resolution 1051 (1996),' 8 October 1997, p. 55.

[18] Ibid.

[19] Ibid, p. 54.

[20] Khadduri, Imad, 'Iraq's Nuclear Non-Capability,' 21 November, 2002.

[21] Hamza, Khidir with Jeff Stein, 'Saddam's Bombmaker, The Terrifying Inside Story of the Iraqi Nuclear and Biological Weapons Agenda,' 2000, p. 334.

[22] Ibid.

2. *Israeli Bombing of Osirak and a SC Resolution*

In October 1981, under orders from, the then Israeli-Prime Minister Yitzhak Rabin, Israeli jets successfully carried out a long-range bombing raid on Osirak at the nuclear centre at Tuwaitha.[23]

The international community condemned the Israeli actions via the SC resolution 487, while the French agreed to rebuild Osirak but eventually they did not.

3. *Role of World Nuclear Suppliers in Developing the Iraqi Nuclear Programme*

As a developing country with a limited technological and industrial base, Iraq depended on external supplies of weapons and production equipment. A readiness among corporations and governments to supply Iraq was the result of relations of trade and war; the relations of monitoring faced by the government of Iraq and its trade partners, which might have hindered such developments, were decidedly weak.[24]

Western suppliers were not interested in politics as much as in profit. On one hand, Iraq, as an oil producing country, had ample money to spend on its nuclear programme and on the other, companies were interested in selling their products to make profit. It was the perfect situation for both and at that time, intelligence services of Western European countries gave these companies a nod to advance in their cooperation with Iraq.[25]

In this regard, it can be concluded that Western intelligence services were turning a blind eye to the activities of Iraq,[26] directly, by allowing shipments of material and equipment to reach Iraq, or indirectly, by not taking any action to halt the movement of such items to Iraq. It was a

---

[23] Bhatia, Shyam and Daniel McGrory, 'Brighter than the Baghdad Sun- Saddam Hussein's Nuclear Threat to the US,' 2000, pp. 135–136.

[24] Deaver, Micheal V., 'Disarming Iraq, Monitoring Power and Resistance,' 2001, p. 35.

[25] Bhatia, Shyam and Daniel McGrory, 'Brighter than the Baghdad Sun- Saddam Hussein's Nuclear Threat to the US,' 2000, p. 159.

[26] Hamza, Khidir with Jeff Stein, 'Saddam's Bombmaker, The Terrifying Inside Story of the Iraqi Nuclear and Biological Weapons Agenda,' 2000, in his book, Hamza stated, "It was a trip to England, however, which exposed me to another dimension of the trade: Behind the scenes, at least in some cases, Western agents seemed to be helping Iraq get what it needed," p. 331.

kind of a mute bargain; Iraq could access the technology it required as long as it was fighting the Mullahs in Iran.

Mr. Jacques Baute confirmed this view, "Intelligence services knew certain things and they turned a blind eye. The IAEA was aware of certain procurements made by Iraq in the early 80s when it was a friend to the West during its war with Iran. All the Iraqi programmes were declared to the Agency. However, at the end of the 80s, they got caught because Iraq became less cautious and the flow of exports slowed down ..."[27]

In this part, the role of different firms in supplying the Iraqi programme with nuclear technology (know-how, materials and equipment) will be discussed.

3.1. *Supplies from German Firms*

German firms played a substantial role in supplying Iraqis with nuclear materials and equipment that were used in the clandestine nuclear programme, as follows:

- The German specialists Boswau and Knauer built Saddam his nuclear sanctuary below the Tigris to withstand any blast.[28]
- A German firm provided Iraq with 84 tons of aluminium alloy tubes—enough to manufacture 6,000 molecular pumps with internally machined grooves designed to keep any uranium hexafluoride leaking through the top bearing from entering the vacuum area between the rotor tube and the outer casing.[29]
- Neue Magdeburger Werkzeugmachinen GmbH provided one of the computer numerically controlled (CNC) machines that had special attachments to manufacture molecular pumps. According to the Eighth Inspection Report, two of the Magdeburger CNC machine tools found in the centrifuge programme were equipped with fixtures designed specifically for machining centrifuge components.[30]

---

[27] Baute, Jacques, Director of Iraq Nuclear Verification Office, in an interview held at the IAEA Headquarters in Vienna, May 2004.

[28] Bhatia, Shyam and Daniel McGrory, 'Brighter than the Baghdad Sun- Saddam Hussein's Nuclear Threat to the US,' 2000, p. 158, and Albright, David and Mark Hibbs, 'Iraq's Shop-Till-You-Drop Nuclear Program,' Bulletin of Atomic Scientist, Volume 42, no. 3, April 1992.

[29] Ibid.

[30] SC, S/23283, 'Note by the UN SG,' transmitting the letter dated from the DG of the IAEA addressed to the SG annexed to it the 'Report on the Eighth IAEA on Site Inspection in Iraq (11–18 November 1991) under the SC Resolution 687 (1991),' 11 December 1991, p. 13.

- A German company supplied about 300 tons of thick aluminium alloy tubes to manufacture outer casings (The rotor spins in a strong, vacuum-tight outer casing to minimise drag and to contain debris if the rotor breaks). An order for 310 tons more was stopped by the 1990 embargo.[31]
- Rhein-Bayern Fahrzeugbau GmbH provided Iraq with 240,000 magnetisable ferrite spacers needed for the ring-shaped stators (hysteresis motors) designed to spin the rotor at high speeds in a vacuum. Rhein-Bayern acted as a middleman for the Iraqi purchase and had design specifications for these "magnets." The spacers and rings were supplied in several shipments to the State Electrical Industries Establishment in Baghdad between January and May 1990, according to the Ninth Inspection Report. From the same company, Iraq acquired a die-casting machine to manufacture coil rings for stators. Iraq also declared it had acquired enough aluminium forgings to manufacture several thousand-bottom flanges, which hold the stator.[32]
- One inspector said that Siemens AG of Germany provided some of the electrical distribution equipment that connected transformers to ion sources at Tarmiya.[33]
- Interatom: This wholly-owned subsidiary of the German firm Siemens concluded a contract in mid-1989 with the Iraqi Industrial Projects Company (IPC) to build a workshop for "tube processing." Western intelligence organisations believed at the time that IPC was a procurement operation for the Iraqi clandestine centrifuge programme. In 1990, Iraqi officials asked Interatom for assistance with the design and production of equipment used in removing enriched uranium gas from the cascade.[34]

According to Interatom documents, the contract included three sections: "training of Iraqi technicians, planning and construction of a workshop complete with appropriate infrastructure, and supply of processing and

[31] Bhatia, Shyam and Daniel McGrory, 'Brighter than the Baghdad Sun- Saddam Hussein's Nuclear Threat to the US,' 2000, p. 158, and Albright, David and Mark Hibbs, 'Iraq's Shop-Till-You-Drop Nuclear Program,' Bulletin of Atomic Scientist, Volume 42, no. 3, April 1992.

[32] Ibid.

[33] Albright, David and Mark Hibbs, 'Iraq's Shop-Till-You-Drop Nuclear Program,' Bulletin of Atomic Scientist, Volume 42, no. 3, April 1992.

[34] Ibid.

testing equipment." Strabag AG, a major German construction firm, was awarded a contract to build the facility.[35]

Germany's export control authority, the Federal Economics Office (known as BAW), approved the first stage of the programme, which was carried out later in 1989. But in mid-1990, after BAW belatedly learnt of the link between Iraqi centrifuge programme and IPC, it cut off the second stage of the programme. On 2 August 1990 (the day Iraq invaded Kuwait), BAW reiterated that "in light of the risk that the training and workshop supply contract would pose to the international reputation of the firm," Interatom should "immediately break off contact" with IPC and not supply equipment to the factory under construction in Iraq. BAW reminded the firm of pending amendments to the German export laws, which would "extend punishment to [companies] that unintentionally, but carelessly, aid nuclear weapons programmes in other countries."[36]

Interatom officials publicly denied that the company had ever contributed any nuclear-related technology or equipment to the Iraqi centrifuge programme and expressed doubt that the facility could have been used in a uranium enrichment programme. But inspectors determined that the workshop in question, code-named B-01, was intended to be an integral part of the Iraqi centrifuge manufacturing programme at Al Furat factory. The Ninth Inspection Report states that upon completion, B-01 would have been used for manufacturing cascade pipe-work and for assembly and mechanical testing of centrifuges, and that one area was possibly intended for a 100-machine cascade.[37]

A sketch of the exterior of B-01 prepared by the IAEA after several inspection visits is almost identical to a sketch of a building on a map given to Siemens personnel to help them find the site. A senior IAEA official said he had no doubt that this map gave directions to Al Furat.[38]

- Iraq had acquired oxidation furnaces from Degussa AG.[39] IAEA officials said Iraq was pursuing the oxidation process but was unsuccessful.[40]

---

[35] Ibid.

[36] Ibid.

[37] Ibid.

[38] Ibid.

[39] SC, S/23283, 'Note by the UN SG,' transmitting the letter dated from the DG of the IAEA addressed to the SG annexed to it the 'Report on the Eighth IAEA on Site Inspection in Iraq (11–18 November 1991) under the SC Resolution 687 (1991),' 11 December 1991, p. 13.

[40] Albright, David and Mark Hibbs, 'Iraq's Shop-Till-You-Drop Nuclear Program,' Bulletin of Atomic Scientist, Volume 42, no. 3, April 1992.

- Reutlinger und Söhne KG provided Iraq with centrifuge balancing machines.[41]
- Electron beam welders were supplied by Leybold Heraeus.[42]
- Arthur Pfeiffer Vakuum Tecknik GmbH provided Iraq vacuum induction furnace (a very high-temperature furnace), which might have been useful in making uranium metal components.[43]
- H&H Metalform supplied several flow-forming machines to Iraq between 1987 and 1989.[44]

The inspectors found seven other H&H flow-forming machines at the Nasser State Establishment for Mechanical Industries, an important support site for the nuclear programme, and two more at a subsidiary establishment at Al Schaula. The Iraqis say these were used to make rocket bodies. The Ninth Report concluded that with appropriate fittings, all of which would have been suitable for making centrifuge rotors.[45]

- Inwako GmbH, a firm located in Bonn and directed by the German arms dealer Simon Heiner, supplied aluminium-nickel-cobalt magnets, which were made by a British firm and procured by another, Endshire Export Marketing, then shipped to Germany before going to Iraq. The supplier of the cobalt-samarium magnets is unknown, although an IAEA inspector said that they might have been made in Britain.[46]

3.2. *Supplies from British Firms*

- Britain learned of Inwako's attempt to help Iraq in 1989, when a British firm solicited by the Germans to supply the magnets alerted export control officials. Britain then tracked effort of Inwako to get the magnets from a second British firm, Endshire Export Marketing, which was willing to supply them. Suspecting that the magnets

---

[41] SC, S/23283, 'Note by the UN SG,' transmitting the letter dated from the DG of the IAEA addressed to the SG annexed to it the 'Report on the Eighth IAEA on Site Inspection in Iraq (11–18 November 1991) under the SC Resolution 687 (1991),' 11 December 1991, p. 13.

[42] Ibid.

[43] Ibid, p. 14.

[44] Ibid, p. 13.

[45] Albright, David and Mark Hibbs, 'Iraq's Shop-Till-You-Drop Nuclear Program,' Bulletin of Atomic Scientist, Volume 42, no. 3, April 1992.

[46] Ibid.

were for the Iraqi nuclear programme, yet, Britain allowed the shipment to proceed. Also, correspondence between Inwako and the British firm Technology Development Group (TDG) shows that TDG had approached the Germans to supply precursor chemicals useful for the production of poison gas.[47]

- TDG: Beginning in 1987, this British firm was co-directed by an Iraqi named Safa Al-Habobi. British investigators believe that under Al-Habobi, TDG and Matrix-Churchill organised the procurement of a wide array of dual- use machine tools, spare parts, and materials for Iraqi military programmes during the late 1980s.[48]
- Matrix-Churchill provided Iraq with CNC machines.[49]

3.3. *Role of Swiss Firms*

- Iraq acquired from the Swiss firm Asea Brown Boveri a large cold isostatic press that could be used to shape explosive charges.[50]
- The Iraqi Electromagnetic Equipment Establishment in Falluja ordered five high-frequency inverters, rated at 15 kilovolts and 25 amperes, from Acomel SA in Neuchâtel, Switzerland, in early 1989. These can be used to power centrifuges.[51]
- Iraqi-owned Durand Properties bought 18% of the Swiss firm Schmiedemeccanica SA, Biasca. The Iraqi procurement firm IPC placed an order with the aforementioned Swiss firm for 1,000 metal forgings. That was intended as the first part of an order for 4,000. The forgings, which look like transmission gears, matched closely the dimensions of forgings that would be inserted in a CNC machine and turned into centrifuge endcaps and baffles. These were confiscated by customs officials at the Frankfurt airport in July 1990. The

---

[47] Ibid.

[48] Ibid.

[49] SC, S/23947, 'Note by UN SG,' annexing a letter dated 12 May 1992 from the Acting DG of the IAEA addressed to the SG, attached to it the 'Report on the Eleventh IAEA on Site Inspection in Iraq (7–15 April 1992) under SC Resolution 687 (1991),' 22 May 1992, p. 14.

[50] SC, S/23283, 'Note by the UN SG,' transmitting the letter dated from the DG of the IAEA addressed to the SG annexed to it the 'Report on the Eighth IAEA on Site Inspection in Iraq (11–18 November 1991) under the SC Resolution 687 (1991),' 11 December 1991, p. 14.

[51] Albright, David and Mark Hibbs, 'Iraq's Shop-Till-You-Drop Nuclear Program,' Bulletin of Atomic Scientist, Volume 42, no. 3, April 1992.

forgings are believed to have been intended for use in CNC machines ordered by Iraq and made by Schäublin AG, another Swiss firm. Five of these machines were seized in 1990.[52]

3.4. *Supplies from American Firms*

- Electron microscopes to take pictures of the barrier pores for the uranium diffusion enrichment process, and advanced computers for nuclear calculations and controls were bought from US firms like IBM.[53]
- Specialised air filters to protect electronics labs were bought from Millipore, a company in Bedford, a suburb northwest of Boston.[54]
- Nupro, an American firm, provided Iraq with uranium Hexafluoride- resistant bellows valves.[55]

3.5. *Supplies from Austrian Firms*

Voest-Alpine AG, an Austrian foundry, has been identified by Western officials as one of several foundries where calutron parts were cast, although the officials do not believe the company knew that the parts were for calutrons. The first order to this foundry came directly from the Iraqi State Electric Establishment. A second order for 28 pieces, six of them identical to the first order, came from a German company. According to the Eighth Inspection Report, the management of the foundry said that the company would have received a larger order from an intermediary if it had been able to meet a required deadline.[56]

---

52 Ibid, and SC, S/24981, 'Note by UN SG,' annexed to it a letter dated 10 December 1992 from the Acting DG of the IAEA addressed to the SG attached to it the 'Report on the Fifteenth IAEA on Site Inspection in Iraq (8–18 November 1992) under SC Resolution 687 (1991),' 17 December 1992, p. 17.

53 SC, S/23283, 'Note by the UN SG,' transmitting the letter dated from the DG of the IAEA addressed to the SG annexed to it the 'Report on the Eighth IAEA on Site Inspection in Iraq (11–18 November 1991) under the SC Resolution 687 (1991),' 11 December 1991.

54 Hamza, Khidir with Jeff Stein, 'Saddam's Bombmaker, The Terrifying Inside Story of the Iraqi Nuclear and Biological Weapons Agenda,' 2000, p. 160.

55 SC, S/23283, 'Note by the UN SG,' transmitting the letter dated from the DG of the IAEA addressed to the SG annexed to it the 'Report on the Eighth IAEA on Site Inspection in Iraq (11–18 November 1991) under the SC Resolution 687 (1991),' 11 December 1991, p. 13.

56 Albright, David and Mark Hibbs, 'Iraq's Shop-Till-You-Drop Nuclear Program,' Bulletin of Atomic Scientist, Volume 42, no. 3, April 1992.

3.6. *Supplies from Yugoslavian Firms*

A Yugoslavian firm, Federal Directorate of Supply and Procurement, may have been the main contractor for the project of building the main calutron site at Tarmiya. Electrical installation was provided by a Yugoslavian company, namely, EMO.[57]

3.7. *Supplies from Japanese Firms*

A Japanese firm, Hamamatsu Photonics KK, legally exported two streak video cameras and support equipment to the Iraqi Ministry of Industry and Minerals in September 1989. Streak cameras are capable of taking high-speed photographs useful in a nuclear weapons development programme.[58]

3.8. *Supplies from Italy, Brazil, Niger*

The three countries provided Iraq with uranium in different combinations, especially uranium dioxide.[59] Italy supplied some machines too.[60]

4. *Tightening Supply Control*

After the revelation of the militarisation of the Iraqi nuclear programme, suppliers realised the need to review their rules to further tighten export controls on dual-use items, especially after the crucial role played by so many companies under their jurisdictions became evident. Mr. Jacques Baute specified that "At the time when Iraq was importing its material, the

---

57 Ibid.

58 SC, S/23283, 'Note by the UN SG,' transmitting the letter dated from the DG of the IAEA addressed to the SG annexed to it the 'Report on the Eighth IAEA on Site Inspection in Iraq (11–18 November 1991) under the SC Resolution 687 (1991),' 11 December 1991, p. 14.

59 SC, S/1997/779, attached to it the 'Letter dated 6 October 1997 from the DG of the IAEA' annexing to it the 'Fourth Consolidated Report of the DG of the IAEA Under Paragraph 16 of the SC Resolution 1051 (1996),' 8 October 1997, p. 25.

60 SC, S/23283, 'Note by the UN SG,' transmitting the letter dated from the DG of the IAEA addressed to the SG annexed to it the 'Report on the Eighth IAEA on Site Inspection in Iraq (11–18 November 1991) under the SC Resolution 687 (1991),' 11 December 1991, p. 40 and Albright, David and Khidhir Hamza, 'Iraq's Reconstitution of Its Nuclear Weapons Program,' Arms Control Today, October 1998.

supply restrictions were lesser than they later became after the discovery of the Iraqi clandestine programme. Iraq was able to get many items from the lists. Hence, these lists needed a revision after the Iraqi case. Consequently, the German authorities, for instance, attempted a stricter control system on the export of these items."[61]

5. *Iraqi Violations of its International Obligations*

As previously mentioned, Iraq had been an NPT party since 1969 and its Safeguards Agreement with the IAEA entered into force at the end of February 1972, which means the IAEA had been carrying out inspections ever since.[62] The Iraqi obligations and violations under both instruments will be discussed:

5.1. *Under the NPT*

The rules of the NPT were previously analysed in Chapter 2 of this study. According to these rules, Iraq has both rights and obligations as an NNWS:

- As an NNWS Iraq, was not supposed to be a recipient of any nuclear weapons, nor to acquire or manufacture them, pursuant to Article II of the Treaty.
- Under Article III, Iraq was allowed to receive nuclear material and equipment for peaceful purposes.
- It was requested to conclude a safeguards agreement with the IAEA to verify fulfilment of its obligations under the Treaty.

Iraq instead decided to pursue a military nuclear programme, which constituted several violations to its obligations under Article II and III of the NPT.

The developments in the Iraqi nuclear programme revealed that Iraq did not receive any nuclear weapons or any other nuclear explosive device directly, but it did receive parts of nuclear explosive devices indirectly (Article II). Iraqi scientists shopped around the world for nuclear materials and equipment to use in their military nuclear programme. Other NPT

[61] Baute, Jacques, Director of Iraq Nuclear Verification Office, in an interview held at the IAEA Headquarters in Vienna, May 2004.

[62] Van den Borre, Tom and Roland Carchon, 'Preventing the Proliferation of Nuclaer Weapons: 50 Years of Atoms for Peace,' Nuclear Law Bulletin No. 57, June 1996, p. 44.

parties who had advanced nuclear technology assisted them. What was the reason behind Iraq's easy access to nuclear technology for its military nuclear program? Was it a loophole in the Treaty that allows the transfer of nuclear technology for peaceful purposes? Was it the permissibility of the dual-use items transfer? Or was it because the implementation of safeguards was weak or not sufficient that it gave way to the diversion of safeguarded materials to military uses?

The political situation during its war with Iran helped Iraq build up its military programmes. Iraq was an ally of the West. The Western Governments permitted their companies to supply Iraq with sensitive nuclear material and equipment needed for its nuclear programme. The NPT provisions allowed such transfers, however, partly due to the contradiction within international law between the sovereignty of the State and rules restricting its behaviour. This is widely exposed in the implementation of IAEA safeguards and the interpretation of its Agreements with NPT member States. The IAEA tries to maintain a balance between the State's sovereignty and its inspecting rights to verify the non-diversion of nuclear material from peaceful to military uses. Because of these sensitivities, implementation of IAEA/NPT safeguards agreements entails several loopholes (as stated in the previous chapters). The rules as they now stand allow States to militarise their nuclear programmes, as long as they do so politically.

Iraq launched a nuclear programme for peaceful purposes during the 70s. The purpose of its nuclear activities changed over the years, eventually concentrating its efforts on building a nuclear explosive device. In that attempt, it received assistance from companies around the world—from Germany, UK, US, France and others—to build its nuclear infrastructure. This constitutes a breach by those States of their NPT obligations. According to Article I of the NPT, NWS should not assist an NNWS to acquire or manufacture nuclear weapons in any capacity whatsoever. The bare fact is what these companies sold to Iraq was in truth assisting it to manufacture a nuclear weapon. Iraq thereby violated its obligations under Article II of the NPT.

In accordance with Article III of the Treaty, Iraq had ratified a Safeguards Agreement with the IAEA for the exclusive purpose of verifying that Iraq was fulfilling its obligations under the NPT with a view of preventing diversion of nuclear energy from peaceful uses to military uses. Procedures for the safeguards were followed with respect to source or special fissionable material, whether it was being produced, processed or used in any principal nuclear facility or even outside any such facility.

Iraq then violated this Article by running a clandestine nuclear programme. The Iraqi Government provided the IAEA with partial declarations about its nuclear equipment and material, although it was under the obligation to declare all its nuclear activities to the Agency to enable the latter to properly implement its safeguards procedures. The IAEA was not aware that Iraq had secret nuclear facilities or that Iraq had brought undeclared nuclear material and equipment into these facilities. Meanwhile, Iraq used declared nuclear facilities to produce undeclared nuclear material.

Iraq was able to abuse some of the failures in the IAEA safeguards system (insufficient declarations, weaknesses of IAEA safeguards techniques, etc.) to serve its interests. By doing so, Iraq was in violation of its obligations under the NPT.

5.2. *Under the Agreement with IAEA for the Application of Safeguards in Connection with the NPT*

Iraq accepted safeguards on all source or special fissionable material in all peaceful nuclear activities for the exclusive purpose of verifying that such material will not be diverted to nuclear weapons or other nuclear explosive devices[63] (Article I of the IAEA Safeguards Agreement). This constituted the Iraqi basic undertaking under its Safeguards Agreement pursuant to paragraph 1 of Article III of the NPT.

Iraq failed to comply with the main objective of safeguards under its IAEA/NPT agreement. According to Article 28, the main objective of safeguards is the timely detection of diversion of significant quantities of nuclear material from peaceful nuclear activities to the manufacture of nuclear weapons or of other nuclear explosive devices or for purposes unknown. As previously stated, Iraq's covert nuclear programme defeated this main objective and succeeded in diverting nuclear material from peaceful activities to the production of nuclear weapons.

In this regard, the DG of the IAEA specified that the technical objectives of the safeguards procedures set forth in the Agreement, as described in Article 28, which relates to the timely detection of diversion of significant quantities of nuclear material, and deterrence of such diversion by means

[63] IAEA, INFCIRC/ 172, 'The Text of the Agreement between Iraq and the Agency for the Application of Safeguards in Connection with the Treaty on the Non-Proliferation of Nuclear Weapons,' entered into force 29 February 1972, and distributed to all IAEA Members on 22 February 1973.

of early detection, does not imply that any minimum quantity of nuclear material is required for the application of safeguards.[64]

Iraq was obligated to cooperate with the Agency to facilitate the implementation of safeguards (Article 3). For years, Iraq cooperated with the Agency by declaring some of its nuclear activities. It always came out clean after the regular IAEA inspections. But, at the same time, Iraq did not facilitate the implementation of safeguards. It did, at times, mislead the inspectors, obstruct their work and conceal information. Even later, while implementing SC Resolutions, Iraqi officials did not sufficiently cooperate with the IAEA inspectors.[65] Iraqi obligations under SC Resolutions did not eliminate its original obligations under existing international binding instruments (NPT and IAEA Safeguards Agreement). Hence, Iraq was still legally committed to follow Article (3).

Article 7(a) of the Agreement states, "Iraq shall establish and maintain a system of accounting for and control of all nuclear material subject to safeguards under this Agreement." Did Iraq establish such a system of control over its nuclear materials? How much did the IAEA know about such a system, if it existed, and how was the information handled by the IAEA during inspections? In fact, throughout the years during which Iraqi facilities fell under safeguards, Iraq had established a system of accounting for control over some, not all, of its nuclear materials. These incomplete records were the ones shown to the IAEA inspectors, during their annual visits, to verify Iraqi declarations on their facilities.

Iraq did not fulfil its commitments under Article 8 (a) of the Safeguards Agreement. In this regard, Iraq allocated only some of its nuclear material to be used in the peaceful nuclear programme, which was properly safeguarded. However, Iraq did not inform the inspectors of the full scope of its nuclear activities at the time.

---

[64] IAEA Board of Governors, GOV/2530, 'A Report by the DG on Non-Compliance by Iraq with its Obligations under the Safeguards Agreement Concluded with the Agency,' Restricted Distribution, 16 July 1991, p. 5.

[65] SC, S/23122, 'Note by the UN SG,' dated 8 October 1991attached to it a letter dated 3 October 1991 from the DG of the IAEA addressed to the SG, Annexed to it the 'First Report on the Sixth IAEA on-Site Inspection in Iraq (22–30 September 1991) under SC Resolution 687 (1991),' 8 October 1991. "The Garage Incident" was clear proof that Iraqi officials tried to obstruct the inspectors' work: they refused to allow the inspectors to carry important documents about the Iraqi programme outside the garage where they had found them, and they locked them in the garage where they had found these documents.

According to the safeguards agreement, Iraq should have accounted for any misuse or loss of nuclear material. However, during the regular IAEA inspections, this did not occur. For one reason or another, Iraq was not declared in violation of its IAEA safeguards agreement for years. It did not declare all its nuclear activities, and according to the IAEA inspection rules, inspectors are not allowed to request further information from a State unless they have substantial information that the State might be concealing information beyond what it had declared to the Agency.

Iraq violated Article (14); it chose to use safeguarded nuclear material in an unsafeguarded nuclear activity without following the procedures necessary to comply with its obligations under this Article. Iraq did not inform the Agency of such diversion. Iraq should have informed the Agency about the total quantity and composition of such unsafeguarded material in Iraq and of any export of such material. Any diversion of such materials should have taken place in accordance with arrangements with the IAEA. Such agreements would not involve any approval or classified knowledge of military activity or relate to the use of the nuclear material therein.

According to the NPT and IAEA Safeguards Agreement with Iraq, the latter had the right to acquire technology relating to enriching uranium or other special fissile material for peaceful purposes only, and IAEA should have been informed that such activities were taking place in Iraq. Such obligations were not honoured. Iraq also concealed that it was involved in chemical reprocessing activities to produce HEU in violation of its Safeguards Agreement. It had seriously embraced a project to divert HEU in Tuwaitha to use it in manufacturing a nuclear bomb in the Tarmiya facility.

The rules here are not adequate to ensure that a State is fulfilling its international legal obligations. Depending solely on the State's declarations leaves a loophole in the rules both theoretically and practically.

The Report of the DG of July 1991 suggested that Iraq had been in non-compliance with its obligations under its Safeguards Agreement with the Agency as follows:

- Article 34(c), which requires nuclear material of a composition and purity suitable for fuel fabrication or isotopic enrichment, and any nuclear material produced at a later stage in the nuclear fuel cycle, which is imported into the State to be subject to all of the safeguards procedures provided for in the Agreement. Article 34 (c) would include, therefore, any domestically produced or imported uranium dioxide of a purity suitable for fuel fabrication, uranium tetrachloride, uranium hexafluoride and enriched uranium, as well as any nuclear material produced there from.[66]

- Article 34(b) applies, inter alia, to imports of yellowcake. It requires that the State should inform the Agency of the quantity and composition of imported materials containing uranium or thorium which has not reached the stage provided for in Article 34(c), unless the material is imported for specifically non-nuclear purposes.[67]
- Iraq did not meet the requirements set forth in Article 42 of the Agreement and the Subsidiary Arrangements[68] General Part concerning submission of design information with respect to facilities.

While clarifying its position with regard to enrichment of uranium activities, the Iraqi Government stated the following:[69]

- The enrichment activities were not subject to safeguards because they were still in the early stages of R&D.
- There was no facility for isotope separation as defined in Article 98 of the Safeguards Agreement.
- The amount of material produced was less than a significant quantity.
- According to Iraqi interpretation, Article 34 (c) referred only to imports of nuclear material of a composition and purity suitable for fuel fabrication, and that as the material in question (a half kilogram of uranium enriched to 4 % U-235) had been produced domestically, Article 34 (c) was not applicable. Alternatively, it was maintained, as the material had not been produced in a "facility" as defined under the Agreement. Article 34 (c) was not applicable.
- Regarding the yellowcake and uranium dioxide referred to in the Iraqi declaration of 7 July, the material was meant for non-nuclear

66 IAEA Board of Governors, GOV/2530, 'A Report by the DG on Non-Compliance by Iraq with its Obligations under the Safeguards Agreement Concluded with the Agency,' Restricted Distribution, 16 July 1991, pp. 3–4.

67 Ibid, pp. 3–4.

68 These are Arrangements concluded between Iraq and the IAEA pursuant to Article 39 of the Safeguards Agreement aimed at detailing the extent necessary to permit the Agency to fulfil its responsibility under the Safeguards Agreement in an effective and efficient manner.

69 IAEA Board of Governors, GOV/2530, 'A Report by the DG on Non-Compliance by Iraq with its Obligations under the Safeguards Agreement Concluded with the Agency,' Restricted Distribution, 16 July 1991, pp. 3–4.

use, and accordingly, the Agency had not been notified of such material in accordance with Article 34 (b) of the Safeguards Agreement.

- With respect to the uranium hexafluoride, the Iraqi view was that the quantity was insignificant and, hence, was not required to be notified or inspected.
- Regarding the uranium tetrachloride, it was maintained that the amount was the normal amount required for feeding experimental separators using the electromagnetic enrichment process and did not require notification.
- Finally, it was noted that the Government of Iraq considered its letter of 7 July 1991 a "corrective" measure meeting the requirements of Article 19 of the Safeguards Agreement.[70]

The difference between the two positions exposes the discrepancy in implementing the rules and procedures of safeguards in an NNWS. The difference in interpretation between what should and should not be safeguarded is very tricky. Here, the judge is the other party, i.e., the IAEA. Although Iraq was not in compliance with its obligations under the Safeguards Agreement, it should nevertheless be inappropriate that the system allows one party of the equation to be a party and judge. It might have been appropriate that the IAEA should have involved the implementation of paragraphs 20–22 of INFCIRC/153 relating to arbitration.[71]

In this case, the DG, as the judge, rightly concluded that the Iraqi Government *failed to report* to the IAEA about the above mentioned nuclear material in accordance with its obligations under the safeguards Agreement. Iraq failed to report to the Agency the existence, on its territory, of nuclear material of the composition and purity provided for in Article 34 (c), including the uranium dioxide, the uranium tetrachloride and the uranium hexafluoride. It failed to report the half kilogram of enriched uranium as nuclear material past the stage provided for in Article 34 (c).[72]

The final judgement was that Iraq had not been in compliance with its obligations under the Safeguards Agreement with the IAEA, in particular with respect to its obligation to accept safeguards on all nuclear material in all peaceful nuclear activities in Iraq.[73]

---

[70] Ibid.
[71] Ibid.
[72] Ibid.
[73] Ibid.

By *concealing information* related to its nuclear activities, Iraq was in breach of its obligations under Articles 37, 49 and 98 of its Safeguards Agreement with the IAEA,[74] as follows:

- The Safeguards Agreement provides no automatic exemption from the application of safeguards to nuclear material which is used or intended for use in peaceful nuclear activities, including research and development activities in the nuclear field.
- The Agreement provides a mechanism for exempting nuclear material of certain quantities (Article 36) and for certain uses (Article 37), both mechanisms require the material to have been subject to safeguards and then, upon request by the State, to be exempted by the Agency.
- The production of nuclear material in a location which is not a "facility" as defined in Article 98 (I) of the Safeguards Agreement does not exempt the State from reporting such material to the IAEA.
- Article 49 of the Agreement requires that when nuclear material is customarily used outside facilities, the State must provide the Agency with, inter alia, a general description of the use of such material, its geographic location and the user's name and address for routine business purposes.
- The State is also obliged to provide a general description of the existing and proposed procedures for nuclear material accountability and control.

These violations remained until they were discovered in 1991. In one of his reports, the DG remarked, "What is clear at this stage is that Iraq's crash programme to extract weapons-grade material from the safeguarded reactor fuel constitutes an additional violation of its safeguards agreement with the IAEA and Iraq's failure, until now, to declare this programme and other weapons related activities and to turn over to the IAEA all nuclear related documents and materials constitutes a violation of Iraq's obligations under the relevant SCR."[75]

74 Ibid, pp. 4–5.

75 SC, S/1995/844, 'Note by the UN SG,' dated 6 October 1995 attached to it a Communication dated 6 October 1995 from the DG of the IAEA addressed to the SG, Annexed to it the 'Eighth report of the DG of the IAEA on the implementation of the Agency's plan for future ongoing monitoring and verification of Iraq's compliance with paragraph 12 of resolution 687 (1991),' Distr. General, 6 October 1995.

Later during 1991, the Iraqi WMD programmes became the subject of several UN resolutions; the Government of Iraq kept refusing the implementation of these resolutions that were calling upon it to disarm. In the reports on the implementation of the IAEA plan for future Ongoing Monitoring and Verification of the Iraqi compliance with SCR 687, the DG of the IAEA, while referring to the nuclear material balance, specified that the Agency's immediate concern was the recent discovery of major inconsistencies between Iraqi declarations with regard to nuclear material and the IAEA analytical results of samples. Specifically, it appeared that the Iraqi authorities may have mischaracterised the origin of certain nuclear material and understated the amounts of material produced and processed in different locations in Iraq or imported from abroad. This gave rise to serious questions about the inventory of nuclear material in Iraq. The Iraqi authorities acknowledged the analytical inconsistencies, but failed to explain the reasons for such inconsistencies refusing to discuss the matter any further. Iraq also refused to provide information to the IAEA inspectors, categorising the questions as too general with the claim that their aim was to maintain the conditions for the embargo.[76]

Iraq refused to recognise its obligations under the relevant SC resolution and that constituted a problem for the IAEA inspectors.[77] As a reaction, the SC condemned Iraqi non-cooperation with the IAEA. In resolutions 707 (1991) and 1115 (1997), the SC condemned Iraq's serious violation of a number of its obligations under several SC resolutions, inter alia, cooperation with the IAEA inspectors. It also condemned the non-compliance of the Government of Iraq with its obligations under its

[76] SC, S/25621, 'Note by the UN SG,' dated 19 April 1993 attached to it a Letter dated 8 April 1993 from the DG of the IAEA addressed to the SG, Annexed to it the 'Third Report of the DG of the IAEA on the implementation of the Agency's plan for future ongoing monitoring and verification of Iraq's compliance with paragraph 12 of resolution 687 (1991),' Distr. General, 19 April 1993, pp. 7–10.

[77] SC, S/26685, 'Note by the UN SG,' dated 3 November 1993 attached to it a Letter dated 29 October 1993 from the DG of the IAEA addressed to the SG, Annexed to it the 'Fourth Report of the DG of the IAEA on the implementation on the Agency's plan for future ongoing monitoring and verification of Iraq's compliance with paragraph 12 of resolution 687 (1991),' Distr. General 3 November 1993, p. 7. The DG stated, "23. Remaining problems are compounded by the persistent refusal of the Iraqi Government to formally recognise its obligations under UN SC resolutions 707 (1991) and 715 (1991). Until Iraq desists in its refusal, it will not have complied with the requirement laid down in paragraph 12 of section C of SC resolution 687 (1991) that it accept the plan prepared by the agency and approved by the SC."

Safeguards Agreement with the IAEA hence constituting a violation of its commitments as a party to the NPT.[78]

To be fair to Iraq, during the attempt to disarm it, the Government of Iraq did not object fully to cooperating with the IAEA solely to challenge the latter's work. Sometimes there were reasonable grounds to Iraqi objections. For example, Iraq objected to accepting inspectors belonging to certain nationalities. The SC denied Iraq this right, in resolution 1137 (1997), it stated, "1. Condemns the continued violations by Iraq of its obligations under the relevant resolutions to cooperate fully and unconditionally with the Special Commission in the fulfilment of its mandate, including its unacceptable decision of 29 October 1997 to seek to impose conditions on cooperation with the Special Commission, its refusal on 30 October 1997 and 2 November 1997 to allow entry to Iraq to two Special Commission officials on the grounds of their nationality, its denial of entry on 3, 4, 5, 6 and 7 November 1997 to sites designated by the Special Commission inspectors on the grounds of their nationality, its implicit threat to the safety of the reconnaissance aircraft operating on behalf of the Special Commission, its removal of significant pieces of dual-use equipment from their previous sites, and its tampering with monitoring cameras of the Special Commission …"[79]

In fact, Robert Gallucci himself, who is an American, and was the Deputy Executive Director of UNSCOM, admitted that he communicated information to the State Department first instead of communicating it to New York, i.e., the UN. He stated, "There were also some issues that arose at that time about how we communicated back to Washington … I had made a very bad mistake in communicating directly to the State Department some information that was supposed to go through New York. I did this because I was not confident about my interlocutor in New York, so I decided to go through the State Department. In terms of being part of an international organisation, this was not a good thing to have done."[80]

---

[78] SC Resolution 707 (1991) adopted by the SC at its 3004th meeting on 15 August 1991. Distr. General, S/Res/707 (1991), Resolution 1115 (1997) adopted by the SC at its 3792nd meeting, on 21 June 1997. Distr. General, S/Res/1115 (1997).

[79] SC Resolution 1137 (1997), adopted by the SC at its 3381st meeting on 12 November 1997, Distribution General, S/Res/1137 (1997).

[80] Gallucci, Robert, 'Reflections on Establishing and Implementing the Post-Gulf War Inspections of Iraq's Weapons of Mass Destruction Programs'—a speech given in a seminar "Understanding the Lessons of Nuclear Inspections and Monitoring in Iraq: A Ten- Year Review, 14–15 June 2001."

So, perhaps the Government of Iraq ought to have had the right to object to cooperating with personnel who break the rules of their work as international officers. The reports concerning the work of the Commission and the IAEA should have been briefed only to the SC (and in the latter's case, to the BG) to decide whether Iraq was in compliance with its international obligations or not and then to decide upon these findings. It was not in the mandate of these officials to transfer any information about Iraq's weapons and capabilities to one or more member States of the UN on a bilateral basis, or because it was their state of nationality.

This was clear in the language of SC resolution 1051 (1996), when it decided that the information provided through the mechanism shall be treated as confidential and restricted to the Special Commission and the IAEA, to the extent that this is consistent with their respective responsibilities under the resolution 715 (1991), other relevant resolutions and the plans for ongoing monitoring and verification approved under resolution 715 (1991).[81]

6. *Process of Disarming Iraq under UN SC Resolutions*

Iraq was the first NPT State to violate the Treaty by implementing a clandestine programme aiming at producing fissile materials and manufacturing a nuclear explosive device.[82]

That was the main reason behind considering the process of disarming Iraq as special. Since 1991, the SC adopted several resolutions to disarm Iraq. There was no other model to pursue in this regard. Hence, a special commission was established to implement the UN resolutions. The nuclear disarmament issue was left to the IAEA because, being the main UN specialised agency concerned with nuclear matters it has the necessary expertise.

In this section, the following will be discussed:

1. Inspection obligations under SC resolutions.
2. IAEA inspectors' rights under SC resolutions.

---

[81] SC Resolution 1051 (1996), adopted by the SC at its 3644th meeting, on 27 March 1996, Distr. General S/RES/1051 (1996), pp. 2–3.

[82] Perricos, Dimitri, 'Understanding the Lessons of Nuclear Inspections and Monitoring in Iraq: A Ten-Year Review'-a Seminar sponsored by the Institute for Science and International Security, 14–15 June 2001.

3. IAEA Inspections to disarm Iraq.
4. Termination of mandates of United Nations Monitoring, Verification and Inspection Committee (UNMOVIC) and Iraq Nuclear Verification Office (INVO).

6.1. *Inspection Obligations under SC Resolutions*

Resolution 687 was endorsed by the SC on 3 April 1991, and was accepted by Iraq. It provided for certain verification tasks. These were (1) to determine the scope and the extent of the nuclear programme; (2) to destroy, remove, or render harmless the components of that programme; and finally (3) to start implementing an Ongoing Monitoring and Verification (OMV) system. These were the parameters of the process from the very beginning.[83]

SCR 687 delegated to the DG of the IAEA the task of implementing the nuclear segment of the disarmament requirements. There were reasons for this. The IAEA already existed as a technical organisation with a lot of experience in areas of inspection. The IAEA DG focused on the organisation to specialise and segregate disarmament from the rest of the Agency. The DG selected a group of IAEA experts who were instructed to carry out activities that differed greatly from those employed during normal inspections. Hence, they carried out surprise inspections and enjoyed free access privileges ("the right to roam") to search for clandestine industrial capabilities related to the nuclear programme. The DG also established a Data Processing and Assessment Centre to make use of the extensive information gained about Iraqi nuclear programme.[84]

States, at the time, needed confidence that the IAEA was sufficiently strong and efficient to carry out any inspection job, especially after it got to be world knowledge that the Iraqi violations took place while Iraq was under IAEA/NPT safeguards and its nuclear facilities were inspected regularly.

What could the Agency offer? It offered experience in organising inspections and inspection campaigns in complex nuclear facilities, and that is what it was expected to do in Iraq. The Agency had experienced staff in nuclear technology. They knew how to take measurements and

---

83 Ibid.

84 Deaver, Micheal V., 'Disarming Iraq, Monitoring Power and Resistance,' 2001, pp. 80–81.

samples of nuclear materials. The Agency had familiarity with high-sensitivity detection technology. It knew how to analyse and evaluate inspection findings.[85]

The SC adopted further resolutions that elaborated specific aspects of the Iraqi disarmament mission, as follows:

- SCR 699 placed the burden of financing the disarmament operations solely upon the Iraqi Government.[86]
- SCR 707 established limitations on the nuclear activities that Iraqi scientists could engage in and demanded that Iraq:[87]
  - Cease immediately any attempt to conceal, move or destroy any material and equipment relating to its nuclear, chemical or biological weapons or ballistic missile programmes, or material or equipment relating to its other nuclear activities, without notification to and prior consent of the Special Commission.
  - Halt all nuclear activities of any kind, except for use of isotopes for medical, agricultural or industrial purposes, until the Council determines that Iraq is in full compliance with its resolutions and the Agency determines that Iraq is in full compliance with its Safeguards Agreement with the Agency.
  - Relinquish all ownership interest in items to be destroyed, removed or rendered harmless pursuant to resolution 687 (1991).
  - The SC asserted a number of rights and privileges for UN personnel implementing disarmament, which Iraqi Government was required to respect, among which were the following:
    - Allow the IAEA inspection team immediate, unconditional and unrestricted access to any and all areas, facilities, equipment, records and means of transportation.

[85] Perricos, Dimitri, 'Understanding the Lessons of Nuclear Inspections and Monitoring in Iraq: A Ten-Year Review'-a Seminar sponsored by the Institute for Science and International Security, 14–15 June 2001.

[86] SC Resolution 699 (1991), adopted by the SC in its 2994th meeting on 17 June 1991.

[87] Perricos, Dimitri, 'Understanding the Lessons of Nuclear Inspections and Monitoring in Iraq: A Ten-Year Review'-a Seminar sponsored by the Institute for Science and International Security, 14–15 June 2001, SC Resolution 707 (1991) stated the rights and privileges given to Iraq and its scientists, and they were further elaborated in the plans for implementation of ongoing monitoring and verification that SCR 715 (1991).

- Make available immediately to the Agency's inspection team any items to which they were previously denied access.
- Allow the Agency's inspection team to conduct both fixed wing and helicopter flights throughout Iraq for all relevant purposes, including inspection, surveillance, aerial surveys, transportation and logistics without interference of any kind.
- Ensure the complete enjoyment, in accordance with its previous undertakings, of the privileges, immunities and facilities accorded to the representatives of the Agency and guarantee their complete safety and freedom of movement.
- Immediately provide or facilitate the provision of any transportation and medical or logistical support requested by the Agency's inspection team.
- Respond fully, completely and promptly to any questions or requests from the Agency's inspection team.

SCR 1051 approved plans for an Export/Import Monitoring Mechanism, it has requested all States to:[88]

- Transmit to the joint unit, constituted by the Special Commission and the DG of the IAEA, the data from potential exporters, and all other relevant information when available to the States, on the intended sale or supply from their territories to Iraq of items or technologies specified in the resolution.
- Report to the joint unit any information they may have at their disposal or may receive from suppliers in their territories of attempts to circumvent the OMV approved by resolution 715 (1991).
- Report to the joint unit if Iraq failed to pursue the procedures for special exceptions to import any specific items laid down by this mechanism.
- Iraq should notify the joint unit of all items and technologies referred to in the mechanism, as of the date agreed upon by Iraq, the Special Commission, and the DG of the IAEA.

The Iraqi Government used several tactics to contest the UN decision among which are the following: (1) concealment of items sought, (2) con-

88 Ibid, pp. 2–3.

cealment of evidence of items sought, (3) redefinition and reconstitution of categorised items, (4) conditional cooperation, (5) harassment, (6) advantageous timing and (7) counterintelligence.[89]

The SC resolutions, especially resolution 1051, addressed the loopholes inherent within the NPT rules. It requested all States, especially potential suppliers, to provide the IAEA and the Special Commission with any relevant data on any intended sale or supply from their territories to Iraq of items and technologies stated in the resolution. This language was new for the suppliers as they were not obligated under the NPT to report to the IAEA about their supplies to NNWS. This, as a result created a loophole in the rules because they only covered cooperation by recipient NNWS.

6.2. IAEA *Inspectors' Rights under* SC *Resolutions*

The IAEA inspectors in Iraq were granted unprecedented rights of access to information, locations, verification technology, and people, with complete logistical support to affect that access. They gathered information and any related data from open sources, from previous inspections, from member States and through interviewing Iraqi scientists. They used different measures and equipment on inspections like surveillance equipment, sensors, seals and optical imagery from satellites, U-2 over-flights, or helicopter over-flights. They verified computer activities and measured nuclear materials where dual-use items were found. They applied new techniques including smear sampling, radiometric sampling, air sampling, and sampling of vegetation and sediment. They also conducted hydrological surveys of the Tigris and Euphrates rivers.[90]

There were no restrictions on IAEA inspectors' movements in the country. They had the right to access any site and any facility they deemed necessary to conduct their inspections. They had the right to confiscate any relevant documents anywhere in Iraq. In short, the inspectors had a free-hand to go anywhere, anytime and gather any information relevant to their work. This differs than the access rights granted to IAEA inspectors under INFCIRC/153 and 540. Mr. Jacques Baute[91] specified that the

---

[89] Deaver, Micheal V., 'Disarming Iraq, Monitoring Power and Resistance,' 2001, p. 91.

[90] Perricos, Dimitri, 'Understanding the Lessons of Nuclear Inspections and Monitoring in Iraq: A Ten-Year Review'-a Seminar sponsored by the Institute for Science and International Security, 14–15 June 2001.

[91] Baute, Jacques, Director of Iraq Nuclear Verification Office, in an interview held at the IAEA Headquarters in Vienna, May 2004.

main difference lay in the right of unconditional access anywhere, anytime, ask anyone, or use any equipment the inspectors wanted to use ... etc. However, even the Model Additional Protocol mentions managed access—not unconditional access. The SC resolutions did not need the country's acceptance to grant the inspectors these wide rights. They were enforced. The resolutions were so stringent that they would never have been approved by States in a protocol.

6.3. IAEA *Inspections to Disarm Iraq*

In the 1980s, the IAEA was performing regular safeguards inspections in Iraq. The Agency was implementing a safeguard system that depended on Iraqi declarations of nuclear material and nuclear activities. Safeguards were not focused on assessing undeclared activities, even if they existed. In 1991, a clandestine nuclear weapons programme was revealed. This was the first time an NPT State violated the Treaty by implementing a clandestine programme to produce fissile materials, to direct its efforts to manufacture a nuclear explosive device.[92]

The effort to disarm Iraq began before Desert Storm, when there were numerous indications that Iraq was building a nuclear weapons programme. A sample which was obtained from Tuwaitha was measured for uranium isotopic revealing its uranium 238 content.[93]

What Really Happened to Iraq's Nuclear Weapon Programme after the 1991 War?

After the 1991 war, the U.N. inspectors were ushered in. Senior Iraqi scientists and engineers in the nuclear cadre were instructed to cooperate with the inspectors. In the first few months, the "clean sheets" were hung up for all to see. Technical documents had been travelling up and down throughout Iraq till an order was issued to return the documents to their original location. David Kay pounced on these documents in the early morning hours of September 1991, and among them were those of Al-Atheer and the bomb specifics.[94]

---

[92] Perricos, Dimitri, 'Understanding the Lessons of Nuclear Inspections and Monitoring in Iraq: A Ten-Year Review'-a Seminar sponsored by the Institute for Science and International Security, 14–15 June 2001.

[93] Nichols, Jere, 'Understanding the Lessons of Nuclear Inspections and Monitoring in Iraq: A Ten-Year Review,' 2001.

[94] Khadduri, Imad, 'Iraq's Nuclear Non-Capability,' 21 November, 2002.

IAEA officials state that poorly timed press leaks accompanied by rumours that Iraq was close to producing bomb quantities of HEU hindered their attempts to gain the confidence of Iraqi scientists and engineers who still withheld important pieces of Iraq's nuclear enigma. However, the main issue was how to get to the bottom of Iraq's nuclear programme.[95]

In 1994, a report surfaced claiming that Iraq was still manufacturing a nuclear bomb and had been working on it since 1991. The IAEA inspectors brought the report to Baghdad, demanding a full explanation.[96]

While implementing resolution 687 Iraq submitted its first declaration on 18 April 1991 which was rejected by the Agency. Iraq denied the existence of any important nuclear material. On 27 April Iraq issued a more detailed declaration which at least gave the Agency a starting point. Iraq declared that it did have HEU as part of the HEU fuel used in the Tamuz reactor. Iraq also admitted that it had buildings and facilities at Tuwaitha other than those that had been visited by safeguards inspectors. On 15 May 1991, the Agency's inspectors performed the first inspection of Iraqi facilities under their new mandate.[97]

Inspections after the Gulf War revealed that in some cases, export controls did not suffice. Iraq's calutron programme was successful largely because few of the materials or components were controlled. Western companies and governments, particularly Germany, do not want to be embarrassed by public revelations about their involvement in the Iraqi nuclear programme. Firms feared that commercial secrets will leak out. But exposing the way Iraq tried to buy itself a nuclear weapons programme is critical to determining where export controls failed—or worked—and what improvements are required if the Iraqi experience is not to be repeated.[98]

As previously mentioned in Chapter 3, Supplier Groups monopolise the nuclear market; their rules go beyond the demands of the NPT/IAEA safeguards. The lack of transparency and cooperation between the NPT Member States and Members of Supplier Groups is behind the loophole

---

[95] Albright, David and Mark Hibbs, 'Iraq's Shop-Till-You-Drop Nuclear Program,' Bulletin of Atomic Scientist, Volume 42, no. 3, April 1992.

[96] Khadduri, Imad, 'Iraq's Nuclear Non-Capability,' 21 November, 2002.

[97] Perricos, Dimitri, 'Understanding the Lessons of Nuclear Inspections and Monitoring in Iraq: A Ten-Year Review'-a Seminar sponsored by the Institute for Science and International Security, 14–15 June 2001.

[98] Albright, David and Mark Hibbs, 'Iraq's shop-till-you-drop nuclear program,' Bulletin of Atomic Scientist, Volume 42, no. 3, April 1992.

in implementation. In the case of Iraq, if Western governments involved in supplying it with sensitive material and equipment had informed the Agency about such supplies, the Agency would have been able to discover the clandestine programme before it went as far as it did. Iraq was in non-compliance with NPT and IAEA safeguards, and the suppliers did not cooperate with the Agency at the time of supply.

The first inspection had a defined objective to get custody of all the nuclear material in Iraq, find out where the material was located, and take control of it so that there would be no danger that the material would disappear. IAEA inspectors found out that Iraq had been accumulating natural uranium since the 80s, both produced in Iraq and procured from abroad. Iraqi scientists were developing the capability to make highly pure uranium metal and to produce plutonium. They developed uranium enrichment capabilities, basically to enrich uranium to weapons-usable levels. They started with gaseous diffusion research, but abandoned it early in 1987. They developed the EMIS programme, which was a dinosaur technology. They were developing gas centrifuge technology and also worked on chemical, laser, and ion exchange technologies. The inspectors found out that, by January 1991, the work by Iraq concentrated on the recovery of uranium at Al Qaim, developing EMIS technology in Tuwaitha and conducting many other development activities in the Tuwaitha area. Although it was badly damaged, they found out that the new R&D area was where Iraq did most of its clandestine development work. They did not find the nuclear material where it was supposed to be, which gave the Iraqis an excuse to hide a lot of things telling the inspectors that they had been lost during the bombing, hidden in the rubble, or burned. During the inspection, the inspectors found out the following:[99]

1. HEU had been removed for storage in the physical protection control bunker at the entrance of Tuwaitha, which the inspectors named "Location A."
2. They were not able to find any spent fuel, except for that which was buried in the rubble at the site. They were finally informed by the Iraqis that they had taken all the spent fuel—one by one, over a long operation, during the bombardment period—and put it in an agricultural farm somewhere closer to Baghdad. In a deserted area, they had dug some holes and put the fuel assemblies in barrels with

[99] Perricos, Dimitri, 'Understanding the Lessons of Nuclear Inspections and Monitoring in Iraq: A Ten-Year Review'-a Seminar sponsored by the Institute for Science and International Security, 14–15 June 2001.

water to be kept there. This they called "Location B." The inspectors had to take the spent fuel under control and store it in areas where it could be later shipped outside the country.

3. In the new R&D area, some of the initial magnets used in the EMIS development programme were found. They were used in the Iraqi enrichment programme.
4. The fuel fabrication laboratory that was being inspected by the Agency under regular safeguards inspections was also visited on this first inspection. There were fresh fuel pins found there in the rubble without contaminating the area.
5. Natural uranium was found kept in a location near Tuwaitha. There were hundreds of barrels of yellow cake. At this particular location, which was called "Location C," the team started to collect all the natural uranium and low-enriched uranium from other locations in Iraq.
6. The inspectors visited a site called Tarmiya that might have had facilities related to gas centrifuge production and development. The team found that, on the whole, the area was in ruins and one thing was very clear: the facility did not have the profile of a centrifuge enrichment facility. But later, they discovered that it was an EMIS facility.[100]
7. The Iraqis constructed a sister facility to Tarmiya in order to develop a greater EMIS enrichment capability and to increase the quantity of material to be enriched. That was done in the area of Ash Sharqat.
8. They had uranium feed material production facility in Al Jesira, also up in the north near Mosul.
9. The gas centrifuge programme was developed with foreign assistance. The programme was initiated at Tuwaitha, but then moved to the Engineering Design Centre (EDC) at Rashdiya. At the same time, while they were doing their R&D work at the EDC, they worked on the construction of a centrifuge production facility in Al Furat. Of course Iraq started thinking of weaponisation, and they concentrated everything in one area, which was the famous Al Atheer site.

The second inspection started with problems, because access was denied at Abu Gharieb facility. It was Maurizio Zifferero who led the inspection

[100] Gallucci, Robert, 'Reflections on Establishing and Implementing the Post-Gulf War Inspections of Iraq's Weapons of Mass Destruction Programs'—a speech given in a seminar 'Understanding the Lessons of Nuclear Inspections and Monitoring in Iraq: A Ten- Year Review,' 14–15 June 2001.

with David Kay there to assist. Zifferero had to return quickly to the SC in New York because of access problems. David Kay proceeded with the rest of the inspection.[101]

The team went over to Falujia, where they found a lot of trucks moving in and out of the facility, despite the requirement of not doing so. It was a standard requirement not to move anything out of a facility once an inspection had been initiated. The inspectors followed the trucks and took pictures of the trucks with Frisbees, while Iraqi guards started shooting at the car of the two IAEA inspectors who were trying to photograph these trucks.[102]

This incident had a positive effect. It promoted more pressure from the SC. It brought about a high-level meeting, which led to more pressure on Iraq to reveal matters, because now it was very obvious that the inspectors had something that could clearly be identified.[103]

The Third Inspection: The inspectors arrived in Baghdad, and on the next morning they were delivered a big list of items that Iraq now wanted to declare. There was a long list of facilities and there were the names of the places where the inspectors could find all these "calutrons" from the EMIS programme. They met with Dr. Jaffar Dia Jaffar, the father of the EMIS programme, who explained what the EMIS programme was all about steadfastly denying that a decision to develop nuclear weapons had been taken or that any weapons development work had been done. The Iraqis showed the inspectors a video of the enrichment work done at Tarmiya. They also provided them with some nuclear material generated by the EMIS process.[104]

But shortly thereafter, Al Atheer came out into the open. Information had come to light which uncovered the famous PC-3 programme. This was the centre of the weapons programme.[105] This facility was visited during the sixth inspection, led by David Kay. The inspectors were detained in a parking lot for four days after discovering the documentation relating to the Iraqi nuclear weaponisation programme. Docu-

---

[101] Ibid.

[102] Ibid.

[103] Ibid.

[104] SC, S/22837, 'Note by the UN SG,' annexed to it a letter dated 25 july 1991 from the DG of the IAEA addressed to the SG, attached to it the 'Third IAEA on Site Inspection Report in Iraq (7–18 July 1991) under SC Resolution 687 (1991),' 25 July 1991.

[105] Perricos, Dimitri, 'Understanding the Lessons of Nuclear Inspections and Monitoring in Iraq: A Ten-Year Review'-a Seminar sponsored by the Institute for Science and International Security, 14–15 June 2001.

mentation seized by inspectors was forcibly confiscated by Iraq and then returned to inspectors after a period of about six hours, after Iraq had removed all documents referring to PC3 Group. This is known as the "Garage Incident".[106]

There were other times where the inspectors had other intelligence information coming in. There were tips of procurement—for example, that Iraq had imported 100 tonnes of maraging steel. The maraging steel was imported in order to make centrifuges. The Iraqis told them that they had melted it down, and therefore they could only provide the melted parts. They were truthful; the inspectors saw the maraging steel slabs, took all the necessary samples, and assessed that the Iraqi declaration was correct in this regard.[107]

End of the Programme

After many discussions, a decision was taken that the buildings that were involved in the programme, as well as the equipment that were used or important for the weapons programme, either had to be taken out or destroyed. All the weapons-usable nuclear material that the inspectors had seen was to be flown out. The first to be flown out of the country was the HEU. The spent fuel was flown out much later, because there were other difficulties involved in that.[108]

The team started with Al Atheer and its famous bunker. To destroy it, they poured several tonnes of concrete until the whole bunker was filled. They brought in special engineers from Sweden and Switzerland to demolish all the buildings that were involved in the programme at Al Atheer, Tarmiya and Ash Sharqat. The calutrons also were completely destroyed.[109]

---

[106] SC, S/23122, 'Note by the UN SG,' dated 8 October 1991attached to it a letter dated 3 October 1991 from the DG of the IAEA addressed to the SG, Annexed to it the 'First Report on the Sixth IAEA on-Site Inspection in Iraq (22–30 September 1991) under SC Resolution 687 (1991),' 8 October 1991, p. 6.

[107] SC, S/23505, 'Note by the SG to the SC,' Annexed to it Letter dated 28 January 1992 from the DG of the IAEA addressed to the SG attached to it the 'Report on the Ninth IAEA on-Site Inspection in Iraq (11–14 January 1992) under SC Resolution 687 (1991),' 30 January 1992, p. 7.

[108] Perricos, Dimitri, 'Understanding the Lessons of Nuclear Inspections and Monitoring in Iraq: A Ten-Year Review'-a Seminar sponsored by the Institute for Science and International Security, 14–15 June 2001.

[109] Ibid.

On 12 January 1992, armed with information from the German Government, Maurizio Zifferero, head of the IAEA's Ninth Inspection mission, told the Iraqis to produce a comprehensive statement about their procurement effort.[110]

The next day, at the Tuwaitha Nuclear Research Centre, technical experts from the Iraqi centrifuge programme gave the inspectors a complete description of the programme to that date. The Iraqis acknowledged that they had imported German materials and components—and added information about acquiring maraging steel and other raw materials needed to manufacture centrifuge components. They claimed that they had destroyed the materials or had made them unusable. As proof, they took inspectors to the State Establishment for Mechanical Works at Iskandariya, 80 kilometres south of Baghdad, and showed them large amounts of powder, remainings of ferrite (magnetisable) spacers for centrifuge motor stators, and flat ingots left after melting down maraging steel. Iraqis also told inspectors that because of tightening export controls, they had made "large procurements as opportunities presented themselves, even though they had no immediate plans for the materials in the quantities ordered." Although, at the time, they were not ready to make large numbers of centrifuges, they knew what they needed to make certain key components.[111]

6.4. *Termination of Mandates of UNMOVIC and INVO*

After lengthy closed debates to terminate the mandate of UNMOVIC and INVO without a final report in accordance with the procedures to terminate the mandate of the subsidiary bodies of the SC, the latter adopted its Resolution no. 1762 on 29th of June 2007[112] which recognised "that the continued operations of UNMOVIC and IAEA's INVO are no longer necessary to verify Iraqi compliance with its obligations under the relevant resolutions" and decided "to terminate immediately the mandates of

[110] SC, S/23505, 'Note by the SG to the SC,' Annexed to it Letter dated 28 January 1992 from the DG of the IAEA addressed to the SG attached to it the 'Report on the Ninth IAEA on-Site Inspection in Iraq (11–14 January 1992) under SC Resolution 687 (1991),' 30 January 1992.

[111] Albright, David and Mark Hibbs, 'Iraq's Shop-Till-You-Drop Nuclear Program,' Bulletin of Atomic Scientist, Volume 42, no. 3, April 1992.

[112] SC Resolution 1762 (2007) adopted by the SC at its 5710th meeting, on 29 June 2007, S/RES/1762 (2007).

UNMOVIC and the IAEA under the relevant resolutions". The same resolution took note of the summary briefings provided by UMOVIC/UNSCOM and IAEA on their respective activities in Iraq since 1991.

Since March 2003, the IAEA was not able to assume its verification role in Iraq.[113] Hence, this step undertaken by the SC was long overdue. It also constituted a precedence and a contradiction to the norm and procedures followed by the UN main organs in terminating any of its subsidiary bodies, especially the SC. The termination of mandate was not based on a final report by UNMOVIC and the IAEA's INVO to reflect the latest developments, achievements to fulfill their mandates and the steps undertaken by the Iraqi Government in order to meet its international legal obligations under the relevant resolutions. Normally, by the end of the mandate of any subsidiary organ established by the SC, the Secretary General provides the latter with a report.[114] The report should indicate that the subsidiary body has fulfilled its mandate; therefore its existence is not necessary. The report should provide the SC with a recommendation to terminate its mandate. This practice has been violated in the case of terminating the mandates of UNMOVIC and INVO.

In this case, political balance was not absent from the SC resolution. The Permanent Members of the SC that led a war against Iraq in 2003, in accurate reports based on allegations that Iraq was still developing WMD and not confining its activities in the peaceful field, were adamant during the negotiations of the resolutions not to receive a final report from UNMOVIC and INVO to end their mandates. Such a report would have exposed to the world, once again and in a UN official document, that the decision to go for the war was based on allegations and not proven facts. From the perspective of these States, this situation was to be avoided even it led to breaking a legal norm in the UN.

For the Iraqi Government, it was just as important to terminate the mandates of UNMOVIC and INVO with or without a final report. Ending a mandate of a political body relieved the Government of Iraq from a body that may constitute constant political pressure. Apart from the political aspect, there were financial reasons. The Iraqi Government

---

[113] SC, S/2005/652, 'Letter dated 13 October 2005 from the Director General of the International Atomic Energy Agency' addressed to the President of the Security Council annexed to a 'Letter dated 14 October 2005 from the Secretary-General addressed to the President of the Security Council,' 17 October 2005.

[114] Repertoire of the practice of the SC, Chapter V: Subsidiary Organs of the SC, volumes covering years from 1946 until 2003.

needed to “transfer the balance in Iraq account opened by the UN to the Development Fund for Iraq, for the purposes of investment”.[115]

On the other hand, it might be worth scrutinising whether the lack of a final report that may have stated that Iraq had no WMD when the Multinational Forces led a war against it and the continuation of the mandates of UNMOVIC and INVO without legal foundation, may have deprived Iraq from its right for compensation for the damages it suffered from caused by implementing SC Resolutions that were based on allegations from receiving refunds for its financial contribution in the work of both bodies from 2003 until 2007.

7. *Conclusions*

The IAEA inspection effort in Iraq provides a snapshot of one country’s success in attempting to procure equipment and know-how for a nuclear weapons programme. Iraq spent 10 years and billions of dollars, but the evidence shows that Iraq got as far as it did because of political circumstances, suppliers’ ignorance, commercial deception, greed and even scientific curiosity. Western governments that backed Iraq against Iran looked the other way when firms sent dual-use items to Baghdad, and companies also justified their behaviour on the basis that Saddam Hussein was battling Teheran’s bearded mullahs. But there is no reason to believe these firms would not have supplied Iran had their governments made another choice, because lax export controls also helped to give Iraq an easy access to certain equipment.[116]

There is, however, little reason to doubt that many firms were victims of Iraqi deception regarding the end uses of their products. Iraq also used intermediaries to conclude contracts with suppliers.[117]

Iraq concealed information from the IAEA in order to retain its military capabilities. The Government of Iraq may also have had mundane reasons related to security and economy concerns for seeking to retain its military capabilities. After implementation of disarmament, Iraq would still have neighbouring States that possessed potent conventional forces

---

[115] SC, S/RES/1762 (2007), Annex II, ‘Letter dated 8 April 2007 from the Minister for Foreign Affairs of Iraq addressed to the President of the Security Council,’ 8 April 2007.

[116] Albright, David and Mark Hibbs, ‘Iraq’s Shop-Till-You-Drop Nuclear Program,’ Bulletin of Atomic Scientist, Volume 42, no. 3, April 1992.

[117] Ibid.

and/or WMD (e.g., nuclear weapons of Israel, chemical weapons of Iran). For reasons of future deterrence and self-defence, officials may have been reluctant to surrender the sought-after capabilities. Another reason may lie in the desire to protect the products of a massive investment in programmes to produce WMD that may have amounted to $ 20 billion.[118]

7.1. *What remained after the inspections?*

The main nuclear centres, that produced weapons-grade uranium like Tuwaitha and Tarmiya, had been destroyed. All that remains of the programme are a few harmless laboratories at the research centre at Tuwaitha.[119]

In the years following 1991, the nuclear weapons project was slowly disbanded. By 1994, its various departments were either elevated to independent civilian industrial enterprises, or absorbed within the Military Industrial Authority.[120]

So, what was left of the programme? At the end of the period—that is to say after the 20 inspections which extended through the middle of 1993—the Agency was able to chart the programme and to start understanding what it was all about. The Agency did not gain a complete understanding, because it took a long time to get additional information. But the team knew, at that time, what they were dealing with; they destroyed or rendered harmless the things that they wanted to destroy, and they knew how and where to disclose of things. So by the middle of the first phase, the team managed to get ample control over what was happening, and how it was happening, in Iraq.[121]

7.2. *Lessons*

There were three points that had to be followed in any inspection, irrespective if the inspection was for very small facilities or very large sites. First, the recording of all findings—there was nothing that was left un-

---

[118] Deaver, Micheal V., 'Disarming Iraq, Monitoring Power and Resistance,' 2001, p. 59.

[119] Bhatia, Shyam and Daniel McGrory, 'Brighter than the Baghdad Sun- Saddam Hussein's Nuclear Threat to the US,' 2000, p. 289.

[120] Khadduri, Imad, 'Iraq's Nuclear Non-Capability,' 21 November, 2002.

[121] Perricos, Dimitri, 'Understanding the Lessons of Nuclear Inspections and Monitoring in Iraq: A Ten-Year Review'-a Seminar sponsored by the Institute for Science and International Security, 14–15 June 2001.

recorded. This was done by the inspectors at the site by hand—writing down all the different events—by taking video recordings, and by photography. Taking photographs was an essential part of each mission. The same was done by the Iraqi side. The second very important point was the follow-up on all of the activities. It was important to see that every inspection was followed up by another inspection, which would build upon the results of the previous one to ensure that there were no gaps left behind. The third important point was that the team took measures to control the movement in and out of the site under inspection. They tried to freeze the situation, so that the Iraqi side would be unable to smuggle out the material.[122]

A few matters might have been handled differently. In the first and second inspections, it might have been possible to obtain better experts—experts who were not solely interested in centrifuge technology, for example. It might have been possible to have a wider range of experts on weaponisation. Another observation is with respect to the method of choosing team members. The method was to select experts with a particular expertise to work for a few days, without really get them to explore, to the extent that they would have wished, the implications of what they thought they were finding. This was illustrated, in several areas, when the inspectors should have been retained for a longer period in order to be more effective.[123]

A second lesson is that the IAEA could dramatically adjust to new circumstances. Unfortunately for the Agency, circumstances required it to adjust yet again only very shortly after going through this experience with Iraq, because of the need for special inspections in North Korea. That change in the IAEA was noticeably interesting. Indeed, the IAEA adjusted, first, in Iraq the way it did, and then, in the North Korean case.[124]

Any inspection regime should create a balance between the "monitoring" function and the "surprise/discovery" function. Somehow, these two have to be in phase. In addition, there has to be a balance struck between

[122] Ibid.

[123] Nichols, Jere, 'Understanding the Lessons of Nuclear Inspections and Monitoring in Iraq: A Ten-Year Review,' 2001.

[124] Gallucci, Robert, 'Reflections on Establishing and Implementing the Post-Gulf War Inspections of Iraq's Weapons of Mass Destruction Programs'—a speech given in a seminar "Understanding the Lessons of Nuclear Inspections and Monitoring in Iraq: A Ten- Year Review, 14–15 June 2001."

the need to get intelligence, and the need to avoid compromising the mission and the organisation in the process of getting the intelligence. This balance is not easy to strike. Also, using modern technology is important. It may be radars, samples analysis or others, for example, in 1991; even usage of GPS was kind of extraordinary. The Iraqis were driven crazy by the inspectors because they used GPS to their way around[125]

Another lesson that should be taken into account is that the IAEA should not cede so many of its rights while negotiating safeguards agreements with Member States. Benjamin Sanders, who stayed for 30 years working in concluding such agreements on behalf of the Agency, clarified, "all through my experience in the IAEA, the Agency's team has had to give in to Member States, otherwise they reached a dead end in the negotiations, and the Agency was all the time the party that was compromising. In my opinion, that was one of the mistakes made with Iraq. According to the NPT, the Agency had more rights than stated in the safeguards agreement with Iraq, which in someway jeopardised the right of the Agency in inspections. For example, the Agency compromised on the number of inspections that it should perform per year. In the safeguards agreement with Iraq, the Agency specified that its inspectors shall perform two inspections per year. However, the NPT left it open. Hence, the IAEA gave its right to initiate further inspections per year other than those specified in the agreement."[126]

To conclude, the Iraqi Government violated all of its international legal obligations. The NPT, IAEA Safeguards Agreement and SCRs were breached.

Although there were regional political circumstances which supported Iraq's development of a clandestine nuclear programme, the country was still in breach of its legal obligations.

Iraq as an NNWS committed to the NPT was in breach of the Treaty provisions when it pursued a military nuclear programme and concealed information regarding its nuclear activities.

It was not complying with its obligations under the IAEA Safeguards Agreement. Iraq violated almost every article that requested it, inter alia, to cooperate with the IAEA to implement the safeguards, in order

---

[125] Ibid.

[126] Sanders, Benjamin, Founder and Executive Chairman of the Programme for Promoting Nuclear Non-Proliferation. Former official, of Netherlands Ministry of Foreign Affairs; IAEA Safeguards Department, and United Nation's Department for Disarmament Affairs. This statement was given in an interview held at UN in 2004.

to report, declare and provide all relevant information related to its nuclear material and equipment. This eventually led to the special process pursued by the SC to disarm Iraq.

Iraq maintained its position of non-cooperation with the IAEA action team, requesting to disarm Iraq, for too long. Some of the Iraqi objections to cooperate with the action team had reasonable grounds. However, Iraq dealt deceitfully with the IAEA inspectors and continued concealing information, which made the latter's job harder.

It is very important to note that the IAEA mandate within the regular inspection process under the IAEA Safeguards Agreement with Iraq was not broad enough to enable the IAEA to discover the Iraqi clandestine nuclear activities. However, the new developments in the IAEA techniques will help in enhancing the efforts of the IAEA to implement its safeguards in a more comprehensive manner and to detect any clandestine activities in Member States in a timely manner. In this regard, Mr Jacques Baute[127] has specified that one of the developments in the IAEA techniques is to look in the open sources and assess the information provided there. If the information provided by media sources, articles, etc. are of relevance and value, the IAEA inspectors are allowed to investigate the credibility of such information with the Member States concerned. This is a new development in the work of inspectors, as they were not previously allowed to do that. They were only allowed to assess the information provided in each country's declaration about its nuclear activities. They were allowed to investigate the information provided in these declarations only—without going through other resources (whether open or not).

The language of the SC resolutions strictly condemned all the Iraqi violations; it was clear in identifying the obligations and requirements the Iraqi Government needed to comply with. The support of the SC was a key component in ensuring the success of the work of both the Special Commission and the IAEA action teams in implementing the Iraqi obligation to disarm. The experience of Iraq proved that if ever another State decided to follow the example of Iraq, the international community would expect the SC to react in the same manner.

However, in the case of Iraq, the violations were evident, but in other cases this might not be the situation. There is a primary discrepancy in the system related to the role of the IAEA in evaluating the sufficiency of

---

[127] Baute, Jacques, Director of Iraq Nuclear Verification Office, in an interview held at the IAEA Headquarters in Vienna, May 2004.

cooperation of a Member State. It plays the role of a party and a judge. The SC, as the political organ of the UN entitled to act upon issues threatening peace and security around the world, receives requests from the IAEA to intervene in issues that may threaten peace and security. However, there is no independent body to judge any difference that may arise between an NNWS and the Agency. If the SC and a UN specialised Agency play the role of a party and a judge, this may jeopardise the impartiality of the regime, (like the example, previously mentioned, of denying Iraq its right to refuse inspectors on a reasonable basis, even though some of them broke the rules of being international officers). These differences should be judged through an Ad Hoc Committee (like arbitration) or they can be raised before the ICJ. This might clear up one of the injustices in the regime.

The situation in the Middle East region and the political circumstances that made it ripe for Iraq to pursue its clandestine activities may rise again and encourage other countries in the region or outside to follow the same path. The development of IAEA inspections will not be solely enough to ensure that there will be no future Iraqs. The IAEA needs more support and continued development of its techniques to ensure the compliance of the NNWS with the NPT and their safeguards agreements with the Agency.

The rules need to develop in order to prevent any more proliferation of nuclear weapons. The IAEA adopted the Model Additional Protocol to Safeguards Agreement (INFCIRC/540) in order to use its inspection rights while conducting its work in NNWS, and at the same time to enable it to use more developed techniques to assist inspectors in the timely detection of any misuse of nuclear material and equipment in the State. However, the Agency is still highly dependant on the cooperation of the NNWS to declare to its inspectors facilities that host peaceful nuclear activities. This is not enough. The IAEA needs the cooperation of all its Member States in this regard.

If the rules of the NPT and the IAEA Safeguards Agreements had covered cooperation by nuclear suppliers with the Agency, that would have assisted the latter in the timely detection of the Iraqi clandestine nuclear programme. For example, if Germany, the UK and the US had informed the Agency of the material and equipment they supplied to Iraq, the Agency would have been able to face the Iraqi authorities with evidence that they were concealing information regarding their nuclear programme in violation of their international obligations.

CHAPTER FIVE

CASE II: IRAN

Iran's nuclear program was inherited from the Shah. He had decided that Iran must acquire nuclear energy. Encouraged by Western countries, he spent around 4 billion dollars to achieve this objective. The pursuit of nuclear energy was effectively suspended after the revolution. But, after the end of the Iraq-Iran War (1980–1988), Iran began to seek external assistance to complete the construction of the Bushehr nuclear reactor, which had been damaged twice by Iraqi air raids during the war.[1]

1. *Nuclear Programme under the Shah*

During the early 1970s, Iran acquired a 5 megawatt nuclear reactor from the US, now called the Amirabad Technical College in Tehran (TRR). The reactor uses a core with 93 % enriched uranium and is regularly inspected by the IAEA. In 1974, the Shah established the Atomic Energy Organisation of Iran and concluded an extendible ten-year nuclear fuel contract with the US, then with Germany in 1976, and with France in 1977. In 1975, he purchased a 10 % share in a Eurodif uranium enrichment plant being built in France. The plant was part of a French, Belgian, Spanish, and Italian consortium. Under the agreement signed by the Shah, Iran was to have full access to the enrichment technology developed by Eurodif, and agreed to buy a quota of enriched uranium from the new plant. By January 1979, Iran had six reactors under contract, and was attempting to purchase a total of 12 nuclear power plants from Germany, France, and the US. Two 1,300 megawatt German nuclear power plants at Bushehr were already 60 % and 75 % completed, and site preparation work had begun on the first of two 935 megawatt French plants at

[1] Roshandel, Jalil, 'Iran, Nuclear Technology and International Security,' The Iranian Journal of International Affairs, Volume 8, Issue 1, 1996, p. 153.

Darkhouin. Those were to be supplied by Framatome. Thousands of Iranians were training in nuclear technology in France, Germany, India, the UK, and the US.[2]

2. *Revitalisation of Iran's Nuclear Programme under Khomeini*

The new Khomeini Government terminated the French and German contracts supporting the program. In March 1979, Iran refused the request of KWU in Germany to mothball the Bushehr reactor projects. As a result, KWU turned the reactor sites over to Iran in late August 1979, and Iran fully abrogated all past agreements with KWU in late November 1979. Also, the scientific cadre was reduced to only 13 people. The Iran-Iraq War, however, soon led the Khomeini Government to revive Iran's nuclear program. With French assistance, Iran established a new nuclear research centre at the University of Esfahan in 1984. This centre was not declared to the IAEA until February 1992, when the IAEA was allowed to make a cursory inspection of six sites that various reports had claimed were the location of Iran's nuclear weapons efforts. Iran announced in 1987 that it had plans to set up a yellow cake plant in Yazd Province. In the mid-1980s, Iran began to show a renewed interest in laser isotope separation (LIS) and opened Jabir Ibn al Hayyan laboratory (JHL) to train Iranian nuclear technicians. Iran also sought foreign support from a range of sources. Pakistan signed a nuclear cooperation agreement with Iran in 1987 under which specialists from the Atomic Energy Organisation of Iran began to train in Pakistan. Iran also strengthened its nuclear research ties with China; they signed a formal nuclear research cooperation agreement in 1990. However, these Iranian efforts suffered major setbacks when Iraq repeatedly bombed Iran's reactor projects at Bushehr.[3]

3. *Developments under Rafsanjani*

The Iranian nuclear programme has been the source of many unconfirmed rumours which exaggerate the size and progress of Iran's effort.[4]

---

[2] Cordesman, Anthony H., 'Iran and Nuclear Weapons'—a Working Draft, Centre for Strategic and International Studies, 7 February, 2000, p. 4.

[3] Ibid, p. 6.

[4] Ibid, p. 7.

In May 1987, Argentina agreed to train Iranian technicians at its Jose Balaseiro Nuclear Institute, and sold Iran $ 5.5 million worth of uranium for its small Amirabad Nuclear Research Centre reactor. In 1991, Iran installed a cyclotron from Ion Beam Applications in Belgium at a facility in Karzaj and signed an agreement with China's Commission on Science, Technology, and Industry for National Defence to build a small 27-kilowatt research reactor at Esfahan to transfer an electromagnetic isotope separator (Calutron) and a smaller nuclear reactor, for "peaceful and commercial" purposes.[5]

Eight different locations have been named as the main Iranian centres of nuclear activities:

1. Tehran: This nuclear research centre consists of a 30-year old, 5 MW research reactor at Tehran University supplied by the US in 1967. Most research activities have been transferred to Esfahan since 1987. This small research reactor does not have the capacity to help Iran produce weapon-grade nuclear material. However, in the 1960s, "Iran acquired from the US some hot cells for the separation of plutonium from spent reactor fuel elements."[6] According to SIPRI's findings, Iran could gain valuable experience in plutonium separation, but it "could not separate enough plutonium for a bomb in this way."[7]
2. Esfahan: No nuclear reactor exists in this nuclear research centre. However, a site has been planned for a 27 MW research reactor to be built by China. The only existing facility there is a small research calutron provided by China in 1987. It should be noted that this small calutron cannot produce the highly enriched plutonium needed to make an atomic bomb. Calutron, as an enrichment technology, is the most expensive and inefficient technology, if compared to other possibilities. Iraq, which used this technology to produce industrial-scale uranium with a maximum of 10% enrichment, had to install about 90 calutrons. It is clear that Iran, by no means, can produce weapon-grade uranium at this site since, as pointed out by the SIPRI report, "Iran does not have unsafeguarded reactors able to produce significant quantities of plutonium."[8]

---

[5] Ibid, pp. 8–9.

[6] Roshandel, Jalil, 'Iran, Nuclear Technology and International Security,' The Iranian Journal of International Affairs, Volume 8, Issue 1, 1996, p. 153.

[7] Ibid, pp. 153–154.

[8] Ibid.

3. Karaj: This facility is known for its medical research centre. In this site there is a cyclotron accelerator acquired from Belgium.[9]
4. Darkhovin: France was expected to construct a nuclear power plant (93.5 MW reactor) in Darkhovin in the 1970s, but the construction never progressed beyond a site survey.[10]
5. Gorgan: Like Darkhovin, this is a planned site for a nuclear power plant (2 × 440 MW reactors) to be built by Russia.[11]
6. Ma'allem Kalayeh: At one point, a 10 MW research reactor was to be constructed by India at this site. This was later cancelled, but the location remains a property of the Atomic Energy Organisation of Iran (AEOI). It is sometimes used as a recreation facility for the staff of the AEOI.[12]
7. Saghand (Yazd): This is a uranium mine and is of no important facility. It was to become operational before 2005.[13]
8. Bushehr: There are two unfinished nuclear power plants (2 × 1200 MW reactors) built by Germany's Kraftwerk Union in the 1970s. These reactors, abandoned after the revolution, have both been damaged in the Iraq-Iran War. The first reactor is up to 80 % complete and the second only about 50 %.[14]

On 8 January 1995, Iran signed an agreement with Russia to complete the reactors at Bushehr.

4. *Proliferation or Non-Proliferation*

It is far easier to know that a nation is proliferating than it is to know specifically where it is proliferating. Some Western specialists, especially British, French, German, US and Israeli experts, are convinced that an Iranian nuclear weapons effort exists; but others, like China and Russia, have effectively rejected this claim when it has been convenient. This contradiction illustrates the fact that there is no international consensus regarding Iran's nuclear activities.[15]

---

[9] Ibid.
[10] Ibid.
[11] Ibid.
[12] Ibid.
[13] Ibid.
[14] Ibid.
[15] Cordesman, Anthony H., 'Iran and Nuclear Weapon'—a Working Draft, Centre for Strategic and International Studies, 7 February, 2000.

5. *The Iranian Breaches, the IAEA Efforts to Ensure Iran's Compliance under its Safeguards Agreement and the Involvement of the SC*

In this context, the following points will be discussed:

1. Iran's obligations under the NPT.
2. Iran's obligations under its safeguards Agreement and the work of the IAEA to ensure the Iranian compliance to its obligations.
3. Iran's Nuclear Programme in the SC.

5.1. *Under the NPT*

Iran signed the NPT on 1 July 1968 and ratified it on 2 February 1970. As an NNWS, Iran has the same legal rights and obligations as Iraq; Iran has the right to develop nuclear capabilities for peaceful purposes, the obligation to accept full scope safeguards on its nuclear facilities and to ensure the non-diversion of nuclear energy from peaceful uses into military ones (Articles II and III).

Iran has been developing its nuclear capabilities for peaceful purposes since the late 50s. It accepted the safeguards of the IAEA in 1973 announcing that its embarking on a long-term plan, based on the merits of energy mix, to construct nuclear power plants with a total capacity of 6,000 MW within two decades.[16]

So far, there is no sign that Iran has diverted any of its nuclear energy from its peaceful program to develop nuclear weapons or any other nuclear explosive device. However, it has enjoyed its right under the NPT to develop its nuclear capabilities for peaceful purposes.

Under the Treaty, Iran can legally build any nuclear facility, including an enrichment plant, as long as the facility is committed to peaceful uses and subject to IAEA safeguards. However, the NPT also allows states to withdraw from the treaty with 90 days notice if their security is fundamentally threatened. Thus, a fully inspected Iran with a uranium enrichment plant is tantamount to Iran having the ability to withdraw from the treaty at any time and rapidly produce weapon-grade material.[17]

---

[16] IAEA, 'Statement by H.E. Reza Aghazadeh, Vice-President of the Islamic Republic of Iran and the President of the Atomic Energy Organisation of Iran at the 46th General Conference of the IAEA,' 16 September 2002, p. 3.

[17] Albright, David & Corey Hinderstein, 'Iran: Furor over Fuel,' Bulletin of Atomic Scientists, May/June 2003, Volume 59, No. 3, p. 12.

These arguments threaten the fundamental principles of the NPT. They hamper the right of NNWS to develop their nuclear capabilities and explore all the nuclear fields for peaceful purposes. NNWS have the right to develop their research and development capabilities to the fullest as long as they do not develop nuclear explosive devices.

As for Iran, the IAEA has not concluded that there is any military aspect to the Iranian nuclear program. A senior Iranian engineer has stated that he did not agree with the hypothesis that a bomb was the only logical outcome of Iran's nuclear quest. He confided that his compatriots were only going for "mastery of the fuel cycle." The capability to make nuclear fuel was one essential ingredient of nuclear and industrial independence. That was not the same as going for a bomb, he argued. Both Germany and Japan, the two defeated powers of World War II, could have independently manufactured nuclear fuels, as they possessed literally tons of plutonium. Either of these powers could produce an atomic bomb within weeks of deciding to do so. Just as Germany or Japan could make nuclear fuels, but had never made the political decision to go for a bomb, so could Iran.[18]

So far Iran has been fulfilling its NPT obligations. However, in recent years, the IAEA has discovered that Iran did not report all aspects relating to its nuclear program. This constitutes a breach by Iran to its NPT obligations. It is committed to accept safeguards and follow its procedures 'with respect to source or special fissionable material whether it is being produced, processed or used in any principal nuclear facility or is outside such facility.' Article III 'shall be applied on all source or special fissionable material in all peaceful nuclear activities within the territory of such a State, under its jurisdiction, or carried out under its control anywhere.'

Acting upon this background, the IAEA inspection teams were involved in an extensive verification process in Iran to ensure the latter's compliance with its obligations. Until now, the Agency has not fund any traces of a military nuclear program in Iran, which is a solace to the international community. Nevertheless, there are still some questions that need to be addressed regarding certain aspects of Iran's nuclear program to ensure that Iran is in full compliance with its international obligations under the NPT and its Safeguards Agreement.

---

[18] Krosney, Herbert, 'Deadly Business—Legal Deals and Outlaw Weapons: The Arming of Iran and Iraq, 1975 to the Present,' 1993. p. 270.

5.2. *Under the Safeguards Agreement with the IAEA*[19]

On the 19th of June 1973, Iran signed a Safeguards Agreement with the IAEA. It also signed the Model Additional Protocol on 18 December 2003. As an NNWS, Iran had been inspected regularly by the IAEA since the entry into force of the Safeguards Agreement. It has been fully complying with its international commitment until several years ago. The IAEA discovered that Iran has invested extensively in the nuclear field without the IAEA being informed of many activities taking place in its territory. Hence, the IAEA inspectors, together with the cooperation of Iran, tried to ensure the latter's compliance with its international obligations under the NPT and the IAEA Safeguards Agreement. In this respect, the following shall be discussed: Iranian international obligations under its Safeguards Agreement, IAEA efforts to ensure the peaceful purposes of Iran's activities and the corrective measures adopted by Iran to facilitate the conduct of the IAEA's mission to ensure Iranian full compliance with its obligations under IAEA safeguards.

Until 2002, the IAEA had not found any suspicious activities within the Iranian nuclear programme. However, as of that date, revelations exposed activities that Iran had concealed for the years from the IAEA. Iran failed to fully comply with its obligations under its Safeguards Agreement, in the following ways:

Iran failed to report to the Agency that it had imported special fissionable material, which didn't fall then under safeguards contrary to the country's obligations under Article 1 of its Safeguards Agreement, which states "The Government of Iran undertakes, pursuant to paragraph 1 of Article III of the Treaty, to accept safeguards, in accordance with the terms of this Agreement, on all source or special fissionable material in all peaceful nuclear activities within its territory, under its jurisdiction or carried out under its control anywhere, for the exclusive purpose of verifying that such material is not diverted to nuclear weapons or other nuclear explosive devices."

According to Article 3 of the Agreement, Iran should have cooperated 'to facilitate the implementation of the safeguards.' At the beginning of the IAEA thorough inspections in Iran in 2002, Iran did not exhibit profound cooperation with the inspection teams. However, since then,

---

[19] IAEA, INFCIRC/212, 'Agreement between Iran and the IAEA for the Application of Safeguards in Connection with the NPT,' signed at Vienna on 19 June 1973.

Iran has acted in compliance with its obligations under Article 18, which states "If the Board, upon report of the Director General, decides that an action by the Government of Iran is essential and urgent in order to ensure verification that nuclear material subject to safeguards under this Agreement is not diverted to nuclear weapons or other nuclear explosive devices, the Board may call upon the Government of Iran to take the required action without delay, irrespective of whether procedures have been invoked pursuant to Article 22 of this Agreement for the settlement of a dispute."

It has cooperated fully with the Agency even beyond its legal obligations under the treaty. It has opened all its sites to inspection and agreed to implement the Additional Protocol to the Safeguards Agreement with the IAEA before its entry into force.[20]

The DG Report of August 2003 stated that "Iran has demonstrated an increased degree of cooperation in relation to the amount and detail of information provided to the Agency and in allowing access requested by the Agency to additional locations and the taking of associated environmental samples. The decision by Iran to start the negotiations with the Agency for the conclusion of an Additional Protocol is also a positive step. However, it should be noted that information and access were at times slow and incremental, and at other times were contradictory to what was previously provided by Iran. In addition, a number of significant issues are still in need of urgent resolution, particularly with regard to Iran's enrichment programme.. Continued and accelerated cooperation and full transparency on the part of Iran are essential for the Agency to be in a position to provide at an early date the assurances required by Member States."[21]

Iran has signed the Model Additional Protocol to the Safeguards Agreement and cooperated with the Agency as stated in the reports of the DG of the IAEA. This proves that the cooperation of a State with the Agency to implement safeguards is crucial. It proves also that the Agency lack of information sources, besides the cooperation of the State itself, affects the system of safeguards negatively. This is in part due to the lack of assistance by other Member States. Supplier States should cooperate

[20] Iran signed the Additional Model Protocol to the IAEA Safeguards on 18 December 2003.

[21] IAEA, GOV/2003/63, DG Report on the 'Implementation of the NPT Safeguards Agreement in the Islamic Republic of Iran,' submitted to the Board of Governors on 26 August 2003, derestricted on 12 September 2003, p. 10.

more with the Agency to ensure that it receives information about where and when special fissionable materials are transported and transferred to the custody of another State.

Article 7 (a) of the Safeguards Agreement between Iran and the Agency requests Iran to establish and maintain a system of accounting for and control of all nuclear material subject to safeguards. Iran has partially fulfilled its obligations under this Article. It has created a system of accounting for some of the nuclear material subject to safeguards; however, it did not mention other sensitive nuclear material from this system that should have been under safeguards. According to Article 32, Iran should have based its system of accounting for and control of all nuclear material subject to safeguards 'on a structure of material balance areas, and shall make provision, as appropriate and specified in the Subsidiary Arrangements, for the establishment of such measures as:

a. A measurement system for the determination of the quantities of nuclear material received, produced, shipped, lost or otherwise removed from inventory, and the quantities on inventory;
b. The evaluation of precision and accuracy of measurements and the estimation of measurement uncertainty;
c. Procedures for identifying, reviewing and evaluating differences in shipper/receiver measurements;
d. Procedures for taking a physical inventory;
e. Procedures for the evaluation of accumulations of unmeasured inventory and unmeasured losses;
f. A system of records and reports showing, for each material balance area, the inventory of nuclear material and the changes in that inventory including receipts into and transfers out of the material balance area.'

As a corrective measure, Iran has submitted to the IAEA accounting reports covering the movements of irradiated targets between ENTC, TRR and JHL.[22]

Iran has concealed information regarding its nuclear material and the features of some of its nuclear facilities. This information was relevant to implementing safeguards and its concealment constituted a breach of Iranian obligations under Article 8 of its Safeguards Agreement. In fact,

---

22 IAEA, GOV/2004/11, DG Report, 'Implementation of the NPT Safeguards Agreement in the Islamic Republic of Iran,' submitted to the Board of Governors of the IAEA on 24 February 2004, derestricted on 13 March 2004, p. 5.

the IAEA has not been informed about the existence of certain material and has never been able to apply safeguards on it. Iran did not follow the procedures stated in Article 14 to use the nuclear material that should be under safeguards in an activity that does not require the application of safeguards.

Iran also did not inform the Agency of the real amounts of nuclear material it possessed for years. It concealed information about the material it had produced and reprocessed in its facilities.

In addition, Iran violated Article 34 b and c; according to these provisions, Iran should have informed the Agency upon importing 'any material containing uranium or thorium' unless the material was 'imported for specifically non-nuclear purposes.' Iran has used such material in nuclear research and development without informing the Agency. Iran has not followed the procedures specified in Article 95, 96 and 97 regarding the transfer of such imported material into Iran. It did not notify the Agency of such transfers; hence, the material was not safeguarded by the IAEA.

The IAEA DG Report of 6 June 2003 confirmed that Iran had not reported to the Agency receiving certain materials. The latter was not aware of these materials except after receiving confirmation by the Supplier State. Later, Iran acknowledged the receipt in 1991 of natural uranium, that had not been reported previously to the Agency, in the form of UF_6 (1000 kg), UF_4 (400 kg) and UO_2 (400 kg). It was stored at the previously undeclared Jabr Ibn Hayan Multipurpose Laboratories (JHL) located at the Tehran Nuclear Research Centre (TNRC). Iran also informed the Agency that, in 2000, it converted most of the UF_4 into uranium metal at JHL. The report verified that the UF_6, UF_4 and UO_2 imported by Iran in 1991 are materials that, as provided for in Article 34(c) of Iran's Safeguards Agreement, should be subjected to safeguards procedures as specified in the Agreement, particularly, the requirement to report inventory changes. Therefore, Iran was obliged to report the import of the material in question at the time of import. Moreover, Iran was obliged to report design information as soon as possible before nuclear material was introduced to the receiving facility, and a Facility Attachment concluded for that facility. Iran also failed to inform the Agency about these materials at the time of their import, however, it confirmed in its letter of 26 February 2003 the receipt of the material in question. According to the Subsidiary Arrangements General Part in force between the Agency and Iran, the latter was supposed to provide the Agency of design information on a new facility no later than 180 days before the introduction of nuclear material into the facility, and the provision of information on a new LOF together with the report relating to the receipt of nuclear material at the LOF. However, as of 26 February 2003, Iran accepted modifications to the Subsidiary Arrangements that the Agency had proposed. As a consequence, Iran is required thereby to inform the Agency of new nuclear facilities and modifications of existing facilities through the provision of preliminary design information as soon as the decision to construct, to authorise construction or to modify has been taken, and to provide the Agency with further design information. Information is to be provided early in the project definition, preliminary design, and construction and commissioning phases.[23]

According to Article 59, Iran should have provided the Agency with detailed reports on nuclear material that was supposed to be safeguarded. However, the Iranian Government did not comply with this provision except at a later stage when the IAEA inspectors were carrying out extensive investigations about the Iranian program.

The same Report of June 2003 identified that Iran had failed to meet its obligations under its Safeguards Agreement with respect to reporting nuclear material, the subsequent processing and use of that material as well as the declaration of facilities where the material was stored and processed. These failures and the actions taken thus far to correct them, can be summarised as follows:[24]

a. Failure to declare the import of natural uranium in 1991 and its subsequent transfer for further processing.

   On 15 April 2003, Iran submitted inventory change reports (ICR) on the import of the UO_2, UF_4 and UF_6. Iran still has to submit ICR on the transfer of the material for further processing and use.

b. Failure to declare the activities involving the subsequent processing and use of the imported natural uranium, including the production and loss of nuclear material, where appropriate, and the production and transfer of waste resulting there from.

   Iran has acknowledged the production of uranium metal, uranyl nitrate, ammonium uranyl carbonate, UO_2 pellets and uranium wastes. Iran must still submit ICR on these inventory changes.

---

[23] IAEA, GOV/2003/40, DG Report, 'Implementation of the NPT Safeguards Agreement in the Islamic Republic of Iran,' submitted to the Board of Governors of the IAEA on 6 June 2003, derestricted on 19 June 2003, pp. 2–6.

[24] Ibid, p. 7.

c. Failure to declare the facilities where such material (including the waste) was received from, stored and processed.

On 5 May 2003, Iran provided preliminary design information for the facility JHL. Iran informed the Agency of the locations where the undeclared processing of the imported natural uranium was conducted (TRR and the Esfahan Nuclear Technology Centre), and provided access to those locations. It provided the Agency access to the waste storage facility at Esfahan, and indicated that access would be provided to Anarak, as well as the waste disposal site at Qom.

d. Failure to provide, in a timely manner, updated design information for the Molybdenum, Iodine and Xenon Radioisotope Production (MIX) Facility and for TRR.

Iran agreed to submit updated design information for the two facilities.

e. Failure to provide information, in a timely manner, on the waste storage at Esfahan and at Anarak.

Iran informed the Agency of the locations where the waste was stored or discarded. It provided the Agency access to the waste storage facility at Esfahan, and indicated that access will be provided to Anarak.[25]

As a corrective measure to these failures, Iran later declared the nuclear material it had imported in 1991. The DG Report of August 2003 identified a number of corrective actions undertaken by Iran which were necessary to enable the Agency to verify the previously unreported nuclear material declared to have been imported by Iran in 1991. These actions included:[26]

a. The submission of ICR on the transfer of the imported UO_2, UF_4 and UF_6 for further processing and use.
b. The submission of ICR on the production of uranium metal, uranyl nitrate, ammonium uranyl carbonate, UO_2 pellets and uranium wastes from the imported material.
c. The provision of design information on the waste storage facility at Esfahan, and granting access to that facility as well as access to Anarak and Qom, where waste resulting from the processing of the imported material was stored or had been disposed of.

---

[25] Ibid.

[26] IAEA, GOV/2003/63, DG Report on the 'Implementation of the NPT Safeguards Agreement in the Islamic Republic of Iran,' submitted to the Board of Governors on 26 August 2003, derestricted on 12 September 2003, pp. 3–4.

d. The submission of updated design information for MIX Facility and for the Tehran Research Reactor (TRR) to reflect activities involving the imported nuclear material.[27]

Iran also failed to declare to the Agency:

a. The pilot enrichment facility at the Kalaye Electric Company workshop; and
b. The laser enrichment plants at Tehran Nuclear Research Centre and the pilot uranium laser enrichment plant at Lashkar Ab'ad.[28]

As a corrective action, Iran submitted ICR relevant to all of these activities, provided design information with respect to the facilities where those activities took place and presented all declared nuclear material for Agency verification. It also undertook, in October 2003, to implement a policy of cooperation and full transparency.[29]

The same gap within the rules exists here. The IAEA failed to exercise its function regarding the implementation of safeguards due to the concealment by Iran of certain information relating to its nuclear programme. This again exposes the inherent contradiction between State sovereignty and the enforcement of international legal rules.

Iran revealed that its nuclear capabilities were more advanced than previously thought—that it was nearly ready to start up a gas centrifuge uranium enrichment pilot plant at Natanz, near Kashan in central Iran. Iran also confirmed that it was building a much larger centrifuge facility at the same site, but insisted that both plants were for peaceful purposes only.[30]

After some delays, Mohamad El Baradei, the DG of the IAEA, visited Iran in February 2003. That visit confirmed that Iran had joined an exclusive club of 10 nations capable of building gas centrifuges.[31]

Due to a loophole in the original IAEA Safeguards Agreement, Iran did not have to declare that it was building a pilot plant until 180 days before it was expected to introduce nuclear material into the plant. It was this loophole that allowed Iran to legally construct the plant

---

[27] Ibid.

[28] IAEA, GOV/2004/83, DG Report on the 'Implementation of the NPT Safeguards Agreement in the Islamic Republic of Iran,' submitted to the Board of Governors on 15 November 2004, p. 19.

[29] Ibid, p. 20.

[30] Albright, David & Corey Hinderstein, 'Iran: Furor over Fuel,' Bulletin of Atomic Scientists, May/June 2003, Volume 59, No. 3, p. 12.

[31] Ibid.

without informing the Agency. In the early 1990s, the IAEA had asked NPT members to close the loophole—to accept a new requirement to provide the Agency with design information when it authorised the construction.[32]

In May 2003, Iran provided preliminary design information on the enrichment facilities under construction in Natanz, which was examined by the Agency. Since March 2003, Agency inspectors have visited facilities at Natanz three times to conduct design information verification and to take environmental samples at the pilot enrichment plant. A first series of environmental and destructive analysis samples has been taken at a number of locations. Additional samples are expected to be taken in the near future. Iran has co-operated with the Agency in this regard. The Agency has presented to the Iranian authorities a safeguards approach for the pilot enrichment plant.[33]

This uranium enrichment program is the most sensitive aspect of the Iranian nuclear program, or in El Baradei's wording 'sophisticated.' The Natanz site houses a centrifuge pilot plant that reportedly has 160 operating centrifuges ready for the introduction of uranium hexafluoride gas. The buildings at Natanz also house centrifuge testing and assembly facilities. A network of companies located elsewhere makes centrifuge components and sends them to Natanz for final testing and assembly. El Baradei and his team also came across enough components at Natanz for another 1,000 machines.[34]

Huge underground buildings were constructed nearby and were built to a depth of 75 feet to withstand aerial attack. When finished, they will hold tens of thousands of centrifuges. When completed (many years from now) these two underground buildings will have a capacity to produce about an estimated 100,000–150,000 separative work units (SWU) per year. This capacity is barely consistent with a program to produce low-enriched uranium for fuel for nuclear power reactors like Bushehr. At this capacity, Natanz could provide about enough low-enriched uranium for one 1,000 megawatt-electric reactor. Iran argues it wants the Natanz facility to establish self-sufficiency in producing nuclear fuel for power reactors, expressing fears of a cutoff of low-enriched uranium by Russia,

---

32 Ibid.

33 IAEA, GOV/2003/40, DG Report, 'Implementation of the NPT Safeguards Agreement in the Islamic Republic of Iran,' submitted to the Board of Governors of the IAEA on 6 June 2003, derestricted on 19 June 2003, p. 6.

34 Albright, David & Corey Hinderstein, 'Iran: Furor over Fuel,' Bulletin of Atomic Scientists, May/June 2003, Volume 59, No. 3, p. 15.

its only supplier. As a result, Iran is proceeding with Natanz, although it must recognise that overseas purchase of low-enriched uranium for fuel is more cost-effective than domestic production. Indigenous production is made even costlier by the perceived need to produce enriched uranium in heavily fortified underground facilities, which are significantly more expensive to build and to operate than aboveground facilities.[35]

The whole problem between the IAEA and Iran is that the latter did not inform the Agency of the aforementioned efforts at the time it was supposed to declare them. There were discrepancies in the information provided by Iran and the findings of the IAEA inspectors. The IAEA DG Report of August 2003, previously mentioned, emphasised that on 11 June 2003, the Agency requested a clarification from the Iranian authorities with regard to the presence of HEU particles at Natanz facility, which was inconsistent with the nuclear material declarations previously provided by the Government of Iran.[36]

On 10 November 2003, Iran informed the DG of the IAEA that it "had decided to suspend all enrichment-related and reprocessing activities in Iran, and specifically to suspend all activities on the site of Natanz, not to produce feed material for enrichment processes and not to import enrichment-related items."[37] Iran later confirmed that the suspension of enrichment activities applied to all facilities in Iran.[38]

This voluntary decision undertaken by Iran represented a confidence-building measure. Nevertheless, the Agency called on Iran to take a further building measure, voluntary to reconsider its decision to begin production testing at the Uranium Conversion Facility and also to reconsider its decision to start construction of a research reactor moderated by heavy water.[39]

There were other pending issues between the Agency and Iran relating to the LEU and HEU contamination found at the Kalaye Electric Company workshop and Natanz.[40] The DG of the IAEA Report of February 2004 stated that it would be very difficult for the Agency to confirm that there

[35] Ibid.

[36] Ibid, p. 2.

[37] IAEA, GOV/2003/75, DG Report, 'Implementation of the NPT Safeguards Agreement in the Islamic Republic of Iran,' submitted to the Board of Governors of the IAEA on 10 November 2003, derestricted on 26 November 2003, p. 4.

[38] IAEA, GOV/2004/60, DG Report, 'Implementation of the NPT Safeguards Agreement in the Islamic Republic of Iran,' submitted to the Board of Governors of the IAEA on 1 September 2004, derestricted on 18 September 2004, p. 9.

[39] Ibid, p. 2.

[40] Ibid, p. 10.

had not been any undeclared nuclear material or activities in Iran, as it was still waiting for Iran to provide requested information detailing the origin of the centrifuge equipment and components, the locations in Iran to which such equipment and components had been moved and the associated details of timescales, and the names of individuals involved. The resolution of this issue will depend to a great extent on the cooperation of the country from which the imported items are believed to have originated.[41]

This is clear evidence that monitoring agencies like the IAEA are incapable of forcing cooperation of the State involved. Hence, this dilemma is a built-in problem. It is engraved in the international regime in general and with disarmament treaties in particular. As long as the enforcement of measures requested by the Agency relies only on the cooperation of the concerned State, the regime will not be able to fulfil its goals in the ideal manner.

Moreover, Iran should have reported to the Agency several incidents relating to contamination and loss of material. Article 68 specifies, "The Government of Iran shall make special reports without delay: (a) If any unusual incident or circumstances lead the Government of Iran to believe that there is or may have been loss of nuclear material that exceeds the limits specified for this purpose in the Subsidiary Arrangements; or (b) If the containment has unexpectedly changed from that specified in the Subsidiary Arrangements to the extent that unauthorised removal of nuclear material has become possible."

In the case of the Kalaye Electric Company workshop, Iran had informed the Agency, when the latter was investigating, that some uranium that had been imported was leaking and hence the walls of the facility were contaminated with traces of LEU and HEU. It had also informed the Agency, in the course of the inspections, that some of the material had been lost during research. Iran should have reported these incidents to the Agency at the time these incidents occurred. There were several contradictions in the Iranian statements with regard to this workshop and the activities that have taken place there. The DG confirmed that during discussions between Iran and the Agency, the latter made reference to information in open sources on the possible conduct of enrichment activities at the workshop of the Kalaye Electric Company in Tehran. The

---

[41] IAEA, GOV/2003/63, DG Report on the 'Implementation of the NPT Safeguards Agreement in the Islamic Republic of Iran,' submitted to the Board of Governors on 26 August 2003, derestricted on 12 September 2003, p. 12.

Iranian authorities acknowledged that the workshop had been used for the production of centrifuge components, but stated that there had been no operations in connection with its centrifuge enrichment development programme involving the use of nuclear material, either at the Kalaye Electric Company or at any other location in Iran. According to the Iranian authorities, all testing had been carried out using simulation studies. A centrifuge component production facility is not a nuclear facility required to be declared to the Agency under Iran's NPT Safeguards Agreement.[42]

However, in his report of August 2003, the DG of the IAEA acknowledged that the results of the environmental samples taken at the workshop revealed particles of HEU—which contradicts the previous statement by Iranian authorities. The latter also stated that these particles must have resulted from contamination originating from centrifuge components which had been imported from outside Iran.[43] Later, Iran acknowledged that "between 1998 and 2002 it had carried out testing of centrifuges at the Kalaye Electric Company using UF_6 imported in 1991; between 1991 and 2000 it had had a laser enrichment programme, in the course of which it had used 30 kg of uranium metal not previously declared to the Agency; and between 1988 and 1992 it had irradiated 7 kilogrammes of UO_2 targets and extracted small quantities of plutonium. Attached to the letter was significant additional information with respect to those activities, as well as information concerning Iran's conversion and heavy water reactor programmes."[44] Iran further acknowledged "on 9 October 2003, that contrary to its previous statements, practically all of the materials important to uranium conversion had been produced in laboratory and bench scale experiments (in kilogramme quantities) between 1981 and 1993 without having been reported to the Agency. These activities were carried out at TNRC and ENTC. The information provided in Iran's letter of 21 October 2003 reveals that, in conducting these experiments, Iran had used nuclear material imported by Iran in 1977 and 1982, some of which had been exempted from safeguards, as well as

---

[42] IAEA, INFCIRC/212, 'Agreement between Iran and the IAEA for the Application of Safeguards in Connection with the NPT,' signed at Vienna on 19 June 1973, p. 2.

[43] IAEA, GOV/2003/63, DG Report on the 'Implementation of the NPT Safeguards Agreement in the Islamic Republic of Iran,' submitted to the Board of Governors on 26 August 2003, derestricted on 12 September 2003, p. 7.

[44] IAEA, GOV/2003/75, DG Report, 'Implementation of the NPT Safeguards Agreement in the Islamic Republic of Iran,' submitted to the Board of Governors of the IAEA on 10 November 2003, and was derestricted on 26 November 2003, p. 4.

safeguarded nuclear material which had been declared to the Agency as a process loss. Iran also declared that, using nuclear material, imported in 1991 and reported to the Agency in February 2003, experiments had been carried out on the conversion of some of the UF_4 to UF_6, and on the conversion of UO_2 to UF_4. On 1 November 2003, Iran agreed to submit all relevant ICR and design information to cover these activities."[45]

Iran has undertaken elaborate measures of concealment to keep its violations of the NPT from being discovered. They have dismantled laser-enrichment equipment and moved it to a hidden storage site and have sanitised a centrifuge workshop at the Kalaye Electric Company before inspectors could check it for evidence of uranium enrichment.[46]

In fact, Iran even started an extensive research and development program on lasers without informing the IAEA about it. However, during the course of 2003–2004 while the IAEA was concluding intensive inspections in Iran, the latter had, as a corrective measure; acknowledged its possession of such a programme. Iran admitted to the Agency's inspectors that it had moved the material and equipment used for this program to another location which the inspectors were allowed to inspect. The IAEA DG Report of November 2003 emphasised that Iran had provided the inspectors with more information on Lashkar Ab'ad and acknowledged that a pilot plant for laser enrichment had been established there in 2000. Iran also stated that uranium laser enrichment experiments had been conducted between October 2002 and January 2003 using previously undeclared natural uranium metal imported from one of the other suppliers. According to Iranian authorities, all of the equipment was dismantled in May 2003 and transferred to Karaj for storage together with the uranium metal. The equipment and material were presented to Agency inspectors at Karaj on 28 October 2003.[47]

So it is misleading to state that the Iranians have dismantled the laser enrichment equipment and moved it to a hidden storage site. The dismantling of the equipment started during the last inspections by the Agency and the Agency was informed that the equipment has been transferred to Karaj site for storage.

---

[45] Ibid, p. 5.

[46] Milhollin, Gary and Valerie Lincy, 'Iran's Nuclear Card,' Commentary Magazine, February 2004, p. 34.

[47] IAEA, GOV/2003/75, DG Report, 'Implementation of the NPT Safeguards Agreement in the Islamic Republic of Iran,' submitted to the Board of Governors of the IAEA on 10 November 2003, derestricted on 26 November 2003, pp. 7–8.

In addition, the IAEA has agreed that the equipment could be further dismantled by Iran and discarded after the results of Agency analysis became available and provided that the results are in agreement with Iran's declarations.[48]

Hence, the IAEA was aware of what was going on with the process in the facility. In his reports, the DG covered any discrepancy between the information provided by the Iranian Authorities and the results of the analysis process of the IAEA investigations stating that the Agency would continue investigating these matters. This discrepancy during the year 2004 occurred with regard to the nuclear material used in the experiments, but not regarding the laser enrichment equipment.

The previous argument does not deny that Iran has concealed ample information from the Agency and that it has been adopting many corrective measures to solve the pending issues with the Agency concerning its nuclear activities. The DG Report of November 2003 emphasised that the recent disclosures by Iran about its nuclear programme showed that, in the past, Iran had concealed many aspects of its nuclear activities, with resultant breaches of its obligation to comply with the provisions of the Safeguards Agreement. Iran's cooperation was limited and reactive, with information slow in coming, changing and contradictory. While most of the breaches involved limited quantities of nuclear material, they have been related to the most sensitive aspects of the nuclear fuel cycle, including enrichment and reprocessing. And although the materials would require further processing before being suitable for weapons purposes, the number of failures by Iran to report the material, facilities and activities in question in a timely manner, as it is obliged to do pursuant to its Safeguards Agreement, has given rise to serious concerns. However, following the Board's adoption of resolution GOV/2003/69, the Government of Iran informed the DG that it had decided to provide the Agency with a comprehensive picture of all its nuclear activities. Since that time, Iran has shown active cooperation and openness. Since then, Iran granted the Agency unrestricted access to all locations the latter requested to visit. It provided information and clarifications in relation to the origin of imported equipment and components and made individuals available for interviews. To date, there is no evidence

[48] IAEA, GOV/2004/11, DG Report, 'Implementation of the NPT Safeguards Agreement in the Islamic Republic of Iran,' submitted to the Board of Governors of the IAEA on 24 February 2004, derestricted on 13 March 2004, p. 2.

that the said undeclared nuclear material and activities were related to a nuclear weapons programme. However, given Iran's past pattern of concealment, it will take some time before the Agency is able to conclude that Iran's nuclear programme is exclusively intended for peaceful purposes.[49]

This reveals, yet again, that cooperation by Supplier States is needed. The lack of transparency in the work of the suppliers negatively affects the monitoring role of the Agency. If the suppliers cooperate with the IAEA, the gap created by the dependency of the IAEA on the State's cooperation may be bridged.

Iranians could attempt to come clean of all of their past illicit activities, bring their country into compliance with the provisions of the NPT, and, within the framework of international law, and work on building a nuclear bomb by legal means. Iran can legally build or import any sort of nuclear plant it requires as long as it can plausibly assert that it is doing so for "peaceful" purposes, and as long as it allows the IAEA to track any nuclear material the plant produces.[50]

The previous arguments are hard to accept as they are, because they lead to a blockade against the transfer of nuclear technology for peaceful purposes to NNWS. It means that even if a country chooses to cooperate extensively with the IAEA to allow the latter to verify and inspect all the facilities in its territory or any other territory under its jurisdiction, this will not be good enough to prevent proliferation. Therefore, the outcome will be hampering the efforts of any NNWS to develop its nuclear capabilities for peaceful purposes. According to Article IV of the NPT each NNWS, including Iran, has the inalienable right "to develop research, production, and use of nuclear energy for peaceful purposes." The treaty also allows "the fullest possible exchange of equipment, materials, and information for the peaceful uses of nuclear energy."

Hence, Iran—with the level of cooperation it has nowadays with the Agency—has the right to "come clean" as a non-proliferator. Iran has taken enormous steps to ensure to the Agency and to the international community that it has been developing its nuclear capabilities for peaceful purposes. It has taken corrective measures towards all the misinfor-

---

[49] IAEA, GOV/2003/75, DG Report, 'Implementation of the NPT Safeguards Agreement in the Islamic Republic of Iran,' submitted to the Board of Governors of the IAEA on 10 November 2003, derestricted on 26 November 2003, p. 10.

[50] Milhollin, Gary and Valerie Lincy, 'Iran's Nuclear Card,' Commentary Magazine, February 2004, p. 34.

mation that it has previously provided to the IAEA. It has further taken voluntary measures to suspend enrichment of uranium activities as a confidence-building measure, although it has affirmed that this does not constitute a legal obligation on Iran[51] so as not to hamper its right to further resume these activities later on. Accordingly, nothing in the NPT, Safeguards Agreement, or Additional Model Protocol to the Safeguards Agreement with IAEA prevents Iran from resuming its efforts to develop its nuclear capabilities for peaceful purposes. It is the job of the IAEA to ensure that there is no diversion of any material from peaceful to military uses. Iran has the right to get into the nuclear field as a consumer and a supplier without violating its obligations under international law.

Again, the IAEA DG Report of November 2003 outlined other failures by Iran to comply with its obligations under the Safeguards Agreement, along with the corrective measures it had undertaken to satisfy those obligations. It also revealed that Iran has "acknowledged that it has been developing, for 18 years, a uranium centrifuge enrichment programme, and, for 12 years, a laser enrichment programme. In that context, Iran has admitted that it produced small amounts of LEU using both centrifuge and laser enrichment processes, and that it had failed to report a large number of conversion, fabrication and irradiation activities involving nuclear material, including the separation of a small amount of plutonium."[52]

Before the IAEA DG report of June 2004 similar issues were raised, specifically two unresolved issues; the first was the failure to determine the origin of the LEU and HEU particles found at Natanz, the Kalaye Electric Company and Farayand Techniques, where Iran has maintained that the contamination originated from imported P-1 centrifuge components. The Agency has received some information about the origin of the components and the contamination that could be useful in resolving this question, but Iran maintains that it does not know the origin of the equipment. It has, however, identified some of the intermediaries involved. The

[51] IAEA, INFCIRC/637, 'Communication dated 26 November 2004 received from the Permanent Representative of France, Germany, the Islamic Republic of Iran and the United Kingdom concerning the Agreement Signed in Paris on 15 November 2004.'

[52] IAEA, GOV/2003/75, DG Report, 'Implementation of the NPT Safeguards Agreement in the Islamic Republic of Iran,' submitted to the Board of Governors of the IAEA on 10 November 2003, and was derestricted on 26 November 2003, pp. 8–9.

other pending issue is the question of the P-2 components that involved a lot of changing or contradictory information submitted by Iran to the IAEA.[53]

The DG, in his statement on 18 June 2004, specified that if these two issues were to be resolved, that would enable the IAEA to report positively on the P-2 programme and to assure the international community that they have inspected all aspects of Iran's nuclear programme.[54]

Since 2004 onwards, the BG continued adopting resolutions in order to imply political pressure on Iran to resolve the outstanding issues relating to its nuclear programme. Although the Board's Resolution GOV/2005/64 dated 11 August 2005 reaffirmed that "the full and sustained implementation of the suspension notified by Iran to the SG on 14 November 2004, as a further voluntary non-legally binding confidence building measure,"[55] it expressed its serious concern that Iran decided to resume the uranium conversion activities in August 2005 and urged Iran to re-establish full suspension of all enrichment related activities. This resolution is a source of contradiction as it implies that the Iranian voluntary decision to halt enrichment activities should be implemented. This also goes beyond the NPT obligations on NNWS since it gives rise to doubts that NNWS should not get engaged in any enrichment related activities.

On the other hand, the BG, in its Resolution GOV/2005/77 dated 24 September 2005,[56] issued a warning to Iran that "the resulting absence of confidence that Iran's nuclear programme is exclusively for peaceful purposes have given rise to questions that are within the competence of the SC, as the organ bearing the main responsibility for the maintenance of international peace and security"[57] and re-urged Iran to suspend its

53 IAEA, GOV/2004/34, DG Report to the Board of Governors, 'Implementation of the NPT Safeguards Agreement in the Islamic Republic of Iran,' GOV/2004/34, 1 June 2004, Derestricted on 18 June 2004, p. 6.

54 IAEA, 'Transcript of the DG's Press Statement on IAEA inspection in Iran, IAEA Headquarters,' 18 June 2004.

55 IAEA, GOV/2005/64, DG Report to the Board of Governors, 'Implementation of the NPT Safeguards Agreement in the Islamic Republic of Iran and related Board Resolutions,' derestricted on 11 August 2005, p. 1.

56 IAEA, GOV/2005/77, DG Report to the Board of Governors, 'Implementation of the NPT Safeguards Agreement in the Islamic Republic of Iran,' derestricted on 24 September 2005.

57 IAEA, GOV/2005/64, DG Report to the Board of Governors, 'Implementation of the NPT Safeguards Agreement in the Islamic Republic of Iran and related Board Resolutions,' derestricted on 11 August 2005, p. 2.

enrichment activities, reconsider the establishment of a research reactor moderated by heavy water and to ratify and implement the Model Additional Protocol. This resolution also went beyond the NPT and IAEA Safeguards' obligations, not only regarding its' request that Iran should suspend its enrichment activities but also to reconsider the establishment of HWR. These are under IAEA verification and safeguards system, as indicated in the DG's consecutive reports since November 2004 onwards.

Moreover, in its resolution GOV/2006/14 dated 4 February 2006, the BG requested the DG to report on the implementation of Iran of its obligations under the BG's resolutions and to convey such reports to the SC. Hence, the monitoring of the implementation of Iran of its obligations under the NPT and IAEA Safeguards moved from a more technical level to a highly politicised one.

5.3. *Iran's Nuclear Programme in the SC*

Based on the BG's resolution GOV/2006/14 of 14 February 2006[58] and the DG report GOV/2006/15 of 27 February 2006,[59] the SC issued its first document regarding Iran's nuclear programme. A Presidential Statement S/PRST/2006/15 was adopted on 29 March 2006.[60] This statement is purely political in nature and constitutes a turning point in the monitoring activities of Iran's nuclear programme. Furthermore, it meant to imply political pressure on Iran, when it requested the DG to report to the SC, within 30 days, the Iranian compliance with the steps requested by the BG.

To fulfil his obligations, the DG submitted his report GOV/2006/27 dated 28 April 2006[61] to the Board and the SC in which he indicated the progress in Iran's compliance with the BG's resolutions. He referred to a letter submitted by Iran on 27 April 2006 that stated, "Iran is fully prepared to continue granting the Agency's inspection in accordance with

[58] IAEA, GOV/2006/14, BG Resolution, 'Implementation of the NPT Safeguards Agreement in the Islamic Republic of Iran,' 4 February 2006.

[59] IAEA, GOV/2006/15, DG Report to the Board of Governors, 'Implementation of the NPT Safeguards Agreement in the Islamic Republic of Iran,' adopted on 27 February 2006, derestricted on 8 March 2006.

[60] SC, S/PRST/2006/15, 'Statement by the President of the SC,' 29 March 2006.

[61] IAEA, GOV/2006/27, DG Report to the Board of Governors, 'Implementation of the NPT Safeguards Agreement in the Islamic Republic of Iran,' GOV/2006/27, on 28 April 2006, derestricted on 12 June 2006.

the Comprehensive Safeguards provided that the Iran's nuclear dossier will remain, in full, in the framework of the IAEA and under its safeguards *and that Iran* is prepared to resolve the remaining outstanding issues reflected in [the DG's] report GOV/2006/15 of 27 February 2006, in accordance with the international laws and norms."[62] He also reiterated that "All the nuclear material declared by Iran to the Agency is accounted for ... However, gaps remain in the Agency's knowledge with respect to the scope and content of Iran's centrifuge programme. Because of this, and other gaps in the Agency's knowledge, including the role of the military in Iran's nuclear programme, the Agency is unable to make progress in its efforts to provide assurance about the absence of undeclared nuclear material and activities in Iran."[63]

The abovementioned letter by Iran reflected its position regarding the interference of the SC in the monitoring activities to its nuclear programmes. It has also indicated its intentions to continue cooperating with the IAEA in accordance with its international legal obligations. Iran has maintained this position until now.

Following these developments, and subsequent to the failure of several attempts to bring Iran, EU, Permanent Members of the SC (P5) and Germany to the negotiating table, the P5 together with Germany and with the support of the High Representative of the EU, transmitted a letter to the SC on 13 July 2006[64] on proposals "for a comprehensive long-term arrangement that would allow for the development of relations and cooperation with the Islamic Republic of Iran." These proposals were based on a package deal to create the right conditions, from the perspective of the proponent States for negotiations with Iran. It has offered incentives to Iran which include suspending the discussion of Iran's nuclear programme in the SC upon the resumption of negotiations, reaffirming the right of Iran to develop its nuclear capabilities for peaceful purposes and committing to support actively the construction of LWRs in Iran through joint international projects. In return for these incentives, they requested Iran to resolve all the outstanding issues with the IAEA, suspend all enrichment related activities and to resume the implementation of the Additional Protocol.[65]

---

[62] Ibid, p. 3.

[63] Ibid, p. 7.

[64] SC, S/2006/521, 'Letter dated 13 July 2006 from the Permanent Representative of France to the UN addressed to the President of the SC,' 13 July 2006.

[65] Ibid, p. 2.

Iran, since 2006 onwards, declared that it considered these proposals as constituting conditions for negotiations and that it will only accept negotiations without any preconditions, as documented in INFCIRC/724.[66]

In an attempt to impose further pressure on Iran, the SC issued its first resolution on Iran's nuclear programme just few days after the aforementioned proposals. SCR S/RES/1696 dated 3 July 2006[67] was another major turning point in the issue. It was adopted under Chapter VII, hence, in accordance with Article 25 of the UN Charter, all Member States of the UN are obligated to accept and carry out such resolution. This resolution demanded Iran to suspend all its enrichment related activities, including research, development and reprocessing. It also called upon all Member States to refrain from transferring to Iran any items, materials, goods and technology related to the mentioned activities. It has endorsed the proposals stated in S/2006/521 and expressed its intention in the case of Iran's non-compliance with its resolution, to adopt any appropriate measures under Article 41 of the Charter. This was a clear threat to Iran that the SC may impose sanctions on it if it concludes that Iran is not in full compliance with its resolution.

Since the adoption of this resolution, the DG submitted two reports to the BG.[68] In these reports, the DG indicated some progress as Iran had provided the Agency access to nuclear material and facilities and the required reports. However, Iran did not provide the Agency with full access to operating records at the Pilot Fuel Enrichment Plant. As for the remaining outstanding issues relating to Iran nuclear activities, no real progress has been achieved.

The story published by Seymour Hersh that the CIA found no conclusive evidence, as yet, of a secret Iranian nuclear-weapons programme running parallel to the civilian operations that Iran has declared to the International Atomic Energy Agency and regardless of such assessment the White House assumes that the lack of evidence means that Iran

---

[66] IAEA, INFCIRC/724, 'Communication dated 26 March 2008 received from the Permanent Mission of the Islamic Republic of Iran to the Agency,' 28 March 2008.

[67] SC Resolution, S/RES/1696 (2006) adopted by the SC in its 5554th Meeting, S/RES/1696 (2006) dated 31 July 2006.

[68] IAEA, GOV/2006/53, DG Report to the Board of Governors, 'Implementation of the NPT Safeguards Agreement in the Islamic Republic of Iran,' on 31August 2006, Derestricted on 14 September 2006, and IAEA, GOV/2006/64, DG Report to the Board of Governors, 'Implementation of the NPT Safeguards Agreement in the Islamic Republic of Iran,' 14 November 2006, derestricted on 23 November 2006.

must have nuclear weapons programme.[69] Furthermore, in a reaction to the previously mentioned DG reports, the SC adopted its Resolution S/RES/1737 dated 27 December 2006,[70] also under Chapter VII of the Charter. This resolution adopted detailed severe measures against Iran. It should be noted that this resolution described the Iranian nuclear activities as of proliferation nature. It went further to request Iran not only to suspend its enrichment related activities but also to suspend work on all heavy water related projects. Describing Iran's nuclear activities of proliferation nature is not based on sound information. Nothing in the DG reports indicated that Iran's activities are directed to military purposes. It has merely reflected that the remaining outstanding issues need further clarification and they are a prerequisite for the Agency to be able to confirm the peaceful nature of Iran's nuclear programme.[71] Also, the SC requested Iran to suspend its enrichment related activities, including research and development thereby contradicting Article IV of the NPT that grants the inalienable right to NNWS for using nuclear energy for peaceful purposes, whether it is related to enrichment activities or to HWR.

Moreover, the SCR imposed targeted sanctions on Iran that included financial, economic sanctions and preventing the supply, sale or transfer of sensitive nuclear materials, goods, equipment and technology to it. It established a list of entities and persons involved in the Iranian nuclear and ballistic missile programmes. It also decided that the IAEA should restrict its provision of technical cooperation to Iran to food, agricultural, medical, safety and humanitarian purposes only. No cooperation should be provided that relates to the, *previously described*, proliferation sensitive nuclear activities. The SC established a Committee to monitor the implementation of this resolution.

It should be noted that according to the UN Charter's provisions, it is solely in the competence of the SC to impose sanctions against any Member State, if it was found in non-compliance with its resolutions. However,

---

[69] Hersh, Seymour M., 'The Next Act: Is a damaged Administration less likely to attack Iran, or more?', 27 November 2006.

[70] SC Resolution, S/RES/1737 (2006), adopted by the SC at its 5612th meeting on 23 December 2006, Distribution General on 27 December 2006.

[71] IAEA, GOV/2006/53, DG Report to the Board of Governors, 'Implementation of the NPT Safeguards Agreement in the Islamic Republic of Iran,' adopted on 31 August 2006, derestricted on 14 September 2006, and IAEA, GOV/2006/64, DG Report to the Board of Governors, 'Implementation of the NPT Safeguards Agreement in the Islamic Republic of Iran,' 14 November 2006, derestricted on 23 November 2006, p. 4.

the restriction of the States' rights to receive technical assistance that are totally safeguarded is contradictory to the NPT and IAEA Statute. Also, the imposition of sanctions by the SC violates principles of human rights and due process in many aspects. The SC, while imposing sanctions relating to freezing assets, confiscating and seizure of funds belonging to individuals or entities, assumes a judicial role and these measures are not subject to any judicial review as well. This is a subject of scrutiny and criticism by the larger membership of the UN. A recent study by the Austrian Ministry of European and International Affairs concluded in collaboration with the Institute for International Law and Justice at New York University School of Law stated in this regard that "Sanctions targeted at individuals have presented a challenge to the authority of the Council: legal proceedings have been commenced in various jurisdictions and there is evidence that sanctions are not always applied rigorously. The Council should be proactive in further improving "fair and clear procedures" to protect the rights of individuals affected by its decisions, complying with minimum standards and providing on its own for periodic review.[72]

Since the adoption of the SCR 1737 (2006), the DG submitted two reports; the first report GOV/2007/7 dated 9 February 2007,[73] reviewed the technical cooperation that the IAEA provides to Iran and decided to continue such cooperation with regard to eleven national projects, twenty regional and two inter-regional projects. However, it has decided not to proceed to provide the said cooperation with regard to four national projects, fourteen regional and four inter-regional projects. The second report, GOV/2007/8 dated 22 February 2007, reflected that some progress has been achieved regarding the provision of access and nuclear materials accounting reports by Iran to the IAEA. Nonetheless, it indicated that the IAEA remained unable to verify the absence of undeclared nuclear material and activities in Iran, unless the latter implements the Additional Protocol and other required transparency measures. It has stated that Iran has not suspended its enrichment related activities

[72] Federal Ministry for European and International Affairs and the Institute for International Law and Justice at New York University School of Law, 'The Role of the Security Council in Strengthening a Rules-Based International System,' Final Report and Recommendations from the Austrian Initiative, 2004–2008, p. i,

[73] IAEA, GOV/2007/7, DG Report to the Board of Governors, 'Cooperation between the Islamic Republic of Iran and the Agency in the light of the UN SCR 1737 (2006),' 9 February 2007, derestricted on 8 March 2007.

and continued construction of the IR-40 Reactor and operation of Heavy Water Production Plant. "In contrast, there has been no indication of reprocessing related activities at any declared sites in Iran."[74]

This previous report initiated the adoption of another SCR, S/RES/1747 dated 24 March 2007[75] that reaffirmed the previous SCR 1696 (2006) and 1737 (2006), clarifying the sanctions the regime imposed against Iran and adding more names of persons and entities to the list established by Res 1737 (2006). This resolution meant to imply further political pressure on Iran, to curb its attempts to develop its nuclear programme and to ensure its cooperation with the IAEA in order to resolve all outstanding issues regarding its nuclear programme to guarantee that it is exclusively for peaceful purposes.

During 2007–2008, the DG submitted several reports to cover the developments on the implementation of Iran of its obligations under the NPT Safeguards Agreement and relevant SCR.

The first report GOV/2007/22 dated 23 May 2007[76] did not reveal anything new compared to the previous reports. However, few months later, on 27 August 2007, Iran and the IAEA reached understandings on the modalities of resolving outstanding issues.[77] These understandings addressed all outstanding issues regarding Iranian nuclear activities aiming at resolving them to ensure that the IAEA has been able to verify the non-diversion of the declared nuclear material at the enrichment facilities in Iran and has therefore concluded that it remains in peaceful use and that the agreement on the outstanding issues "shall further promote the efficiency of the implementation of safeguards in Iran and the IAEA's ability to conclude the exclusive peaceful nature of Iran's nuclear activities."[78]

---

[74] IAEA, GOV/2007/8, DG Report to the Board of Governors, 'Implementation of the NPT Safeguards Agreement and Relevant Provisions of SCR 1737 (2006) in the Islamic Republic of Iran,' 22 February 2007, derestricted on 7 March 2007.

[75] SC Resolution, S/RES/1747 adopted by the Security Council at its 5647th meeting, dated 24 March 2007.

[76] IAEA, GOV/2007/22, DG Report to the Board of Governors, 'Implementation of the NPT Safeguards Agreement and Relevant Provisions of SC Resolutions in the Islamic Republic of Iran,' 23 May 2007.

[77] IAEA, INFCIRC/711, 'Communication dated 27 August 2007 from the Permanent Mission of the Islamic Republic of Iran to the Agency concerning the text of the "Understandings of the Islamic Republic of Iran and the IAEA on the Modalities of Resolution of the Outstanding Issues,' 27 August 2007.

[78] Ibid.

As of August 2007, the DG reports reflected the resolution of several outstanding issues regarding Iran's nuclear programme, including IAEA's conduct of design information verification at IR-40 Reactor, the resolution of the questions regarding Plutonium separation experiments carried out at TRR which ensured that Iran's statements concerning these experiments are consistent with IAEA's findings and concluded that the statements of Iran relating to the source of the irradiated HEU in the coolant system and irradiated HEU contamination are not inconsistent with the Agency's findings. It has considered this question as no longer outstanding.[79] IAEA has also concluded that the physical inventory of the nuclear material at the uranium conversion facility as declared by Iran was consistent with the results it assessed of the physical inventory verification.[80] Moreover, the Agency's inspectors were able to carry out unannounced inspections[81] and Iran has provided sufficient access to individuals and has responded in a timely manner to questions and provided clarifications and amplifications on issues raised in the context of the work plan.[82] IAEA concluded that Iran's answers regarding past P-1 and P-2 centrifuge programmes are consistent with its findings.[83] It considered that the question of the Polonium-210 experiment is no longer outstanding after concluding that all information provided by Iran is consistent with its findings.[84] It has also reached the same conclusion with regard to the Gchine Machine issue.[85]

Certain activities were conducted under the Agency's containment and surveillance, including the current enrichment related activities in

[79] IAEA, GOV/2008/4, DG Report to the Board of Governors, 'Implementation of the NPT Safeguards Agreement and relevant provisions of Security Council resolutions 1737 (2006) and 1747 (2007) in the Islamic Republic of Iran,' 22 February 2008, p. 3.

[80] IAEA GOV/2007/48, DG Report to the Board of Governors, 'Implementation of the NPT Safeguards Agreement in the Islamic Republic of Iran,' 3 August 2007, derestricted on 12 September 2007, p. 3.

[81] IAEA, GOV/2008/4, DG Report to the Board of Governors, 'Implementation of the NPT Safeguards Agreement and relevant provisions of Security Council resolutions 1737 (2006) and 1747 (2007) in the Islamic Republic of Iran,' 22 February 2008, p. 3.

[82] IAEA, GOV/2007/58, DG Report to the Board of Governors, 'Implementation of the NPT Safeguards Agreement and relevant provisions of Security Council resolutions 1737 (2006) and 1747 (2007) in the Islamic Republic of Iran,' 15 November 2007.

[83] Ibid, p. 3.

[84] IAEA GOV/2007/48, DG Report to the Board of Governors, 'Implementation of the NPT Safeguards Agreement in the Islamic Republic of Iran,' 3 August 2007, derestricted on 12 September 2007, p. 9.

[85] Ibid.

Iran. The Agency was able to confirm that no reprocessing activities have been carried out with regard to the declared facilities by Iran (TRR, MIX facility and IR-40).[86] "The one major outstanding issue relevant to the nature of Iran's nuclear programme is the alleged study on the Green Salt Project, high explosive testing and the missile re-entry vehicle." This is a matter "critical to an assessment of possible military dimension to Iran nuclear programme."[87] The Agency provided Iran with additional documentation on the alleged studies on 15 February 2008 and the latter has maintained its position that these allegations are baseless.[88] Furthermore, the IAEA "has note detected the use of nuclear material in connection with the alleged studies issue, the Agency has no concrete information about possible current undeclared nuclear material and activities in Iran."[89] However, to provide "credible assurances about the absence of undeclared nuclear material and activities in Iran, the Agency needs to reach some clarity about the nature of the alleged studies and that Iran implements the Additional Protocol."[90]

Based on these developments, the SC adopted a new resolution S/RES/1803 dated 3 March 2008[91] in which it welcomed the agreement between Iran and the IAEA to resolve all outstanding issues concerning Iran's nuclear programme, reaffirmed the previous SCR and updated the list of individuals and entities involved in nuclear and ballistic missile programmes.

Following the adoption of this resolution the P5 and Germany, with the support of the High Representative of the EU communicated a statement deploring the continued failure of Iran to comply with its SC and IAEA Board requirements, in particular by expanding its enrichment related activities. It noted the progress made in implementing IAEA-Iran work plan and "IAEA's concerns about the "alleged studies," which are critical to an assessment of a possible military dimension to Iran's nuclear

---

86 Ibid, p. 8.

87 Ibid, p. 6.

88 IAEA, INFCIRC/724, 'Communication dated 26 March 2008 received from the Permanent Mission of the Islamic Republic of Iran to the Agency,' 28 March 2008, p. 2.

89 IAEA GOV/2007/48, DG Report to the Board of Governors, 'Implementation of the NPT Safeguards Agreement in the Islamic Republic of Iran,' 3 August 2007, derestricted on 12 September 2007, p. 10.

90 Ibid.

91 SC Resolution, S/RES/1803, adopted by the Security Council at its 5848th Meeting, on 3 March 2008.

programme."[92] It called upon Iran to heed the said requirements and the communicating States expressed their commitment to a negotiated solution with Iran reaffirming their proposals presented to Iran in June 2006. They have delegated, the EU's High Representative for the Common Foreign and Security Policy, Javier Solana, to address the interests and concerns of both sides with the Secretary of Iran's Supreme National Security, Saeed Jalili in order to "gradually create the conditions for the opening of negotiations."[93] This Statement is purely political in nature, but it implies further pressure on Iran to resume negotiations and to suspend its enrichment related activities.

Iran probably perceived the increasing political pressure by the aforementioned powers and the continuity of sanctions imposed against it by SCR, as unfair and beyond Iran's obligations as an NNWS under the NPT and its Safeguards Agreement with the IAEA. Furthermore, it does not take into real consideration the progress achieved in the level of cooperation between Iran and the IAEA. As a reaction, it has communicated a Note Verbale dated 26 March 2008 to the DG expressing its observations with respect to the SCR 1803 (2008).[94] Iran, in this communication, detailed its understanding to the SCR. It considered the engagement of the SC in the Iranian programme unlawful and failed to follow the appropriate procedures required for its intervention. The SCR are in contradiction with the UN Charter and international law and exceed the SC's mandate that constitutes an obvious instance of *ultra vires*, especially that the SC adopts enforcement measures against a legitimate conduct by a State. Such a conduct shall not be introduced as an instance of threat to international peace and security especially in the absence of any violation of the NPT and IAEA Statute by Iran.[95] In this regard, Iran quoted the International Criminal Tribunal for the Former Yugoslavia (ICTY) that "the determination that there exists a threat is not a totally unfettered discretion, as it has to remain, at the very least, within the limits of the purposes and principles of the Charter."[96] Hence, Iran declared that it is not obliged

[92] IAEA, INFCIRC/723, 'Communication dated 4 March from the Governor for the Russian Federation and the Resident Representatives of China, France, Germany, the United Kingdom and the United States of America concerning UN SC resolution 1803 (2008),' 5 March 2008.

[93] Ibid, p. 4.

[94] IAEA, INFCIRC/724, 'Communication dated 26 March 2008 received from the Permanent Mission of the Islamic Republic of Iran to the Agency,' 28 March 2008, p. 1.

[95] Ibid, p. 12.

[96] Ibid.

to implement the unlawful demands of the SC and it does not consider the SCR as those covered by Article 25 of the Charter.[97]

The Iranian communication is based partially on international law and partially on political rhetoric Statements. Iran rightly indicated that the SCR contradicts the NPT, namely Article IV and the IAEA Statute regarding the rights of NNWS to develop its nuclear capabilities for peaceful purposes. However, even if the procedures required by the IAEA Statute and the Safeguards Agreement between the IAEA and Iran may not have been appropriately followed, the fact remains that there are pending issues to be resolved between the IAEA and Iran. The DG reports verified that the declared nuclear materials and activities in Iran have not been diverted towards military purposes, and it has also repeatedly indicated that the Agency is not in a position to "make progress towards providing credible assurances about the absence of undeclared nuclear material and activities in Iran before reaching ample clarity about the nature of the alleged studies, and without the implementation of the Additional Protocol."[98]

It should be noted that the legal debate that Iran brought up before the international community in INFCIRC/724 regarding the implementation of Article 25 of the Charter, and how far Member States are obliged to accept and carry out the SCR if they are deemed illegal or *ultra vires* is in need to be addressed. The SC has been criticised by the larger membership of the UN because it has frequently encroached on the GA and ECOSOC mandates. Although, the issues of mandates should be dealt with in the GA, the issue of the illegality of SCR and the obligations of Member States under such resolutions might be the basis to initiate a request by the GA for an advisory opinion from the ICJ on the matter. The ICJ is not mandated to review the legality of the SC resolutions as such, because in accordance with the Charter, both are main organs of the UN and the Charter did not grant the Court such authority. However, the previous question regarding the obligations of Member States to accept and carry out illegal SCR falls within the competence of the ICJ, hence, this question can and should be addressed.

On 26 May 2008, the DG submitted another report relating to Iran's nuclear programme[99] in which he indicated that Iran has agreed to

[97] Ibid, p. 17.

[98] Ibid.

[99] IAEA, GOV/2008/15, DG Report to the Board of Governors, 'Implementation of the NPT Safeguards Agreement and relevant provisions of Security Council resolutions 1737 (2006), 1747 (2007) and 1803 (2008) in the Islamic Republic of Iran,' 26 May 2008.

address the alleged studies on the Green Salt Project, high explosives testing and the missile re-entry vehicle project. However, Iran maintains that these allegations are baseless and that the data has been fabricated.[100] The DG stated that "substantive explanations are required from Iran to support its statement on the alleged studies, including the role of the uranium metal document, some procurement activities of Iran's military related institution and on other information with a possible military dimension."[101] In this context, he has also indicated that "the Agency has not detected the actual use of nuclear material in connection to the alleged studies."[102]

The report also mentioned that Iran did not act in accordance with the SCR, which is similar to Iran's declaration in INFCRIC/724 that it is not bound by them. Iran has not implemented the modified text of its Subsidiary Arrangements General Part, Code 3.1 on the early provision of the design information, has not suspended its enrichment related activities and has continued the construction of IR-40.

On 18 June 2008, Iran decided to propose a new package for constructive negotiations. This package has several dimensions. It includes political, security, economic and nuclear related issues. In the view of Iran, the package has to be comprehensive to establish long-term cooperation between the parties and contributes to the sustainability and strength of regional and international security and just peace.[103]

Regarding the nuclear issue Iran is ready to consider the following:

1. Obtaining further assurance about the non-diversion of nuclear activities of different countries.
2. Establishing enrichment and nuclear fuel production consortiums in different parts of the world—including Iran.
3. Cooperation to access and utilise peaceful nuclear technology and facilitating its usage by all states.
4. Nuclear disarmament and establishment of a follow up committee.
5. Improved supervision by the IAEA over nuclear activity of different states.
6. Joint collaboration over nuclear safety and physical protection.

---

[100] Ibid, p. 5.

[101] Ibid.

[102] Ibid.

[103] IAEA, INFCIRC/729, 'Communication dated 16 June 2008 received from the Permanent Mission of the Islamic Republic of Iran to the Agency concerning the text of the Islamic Republic of Iran's proposed package for constructive negotiation,' 18 June 2008.

7. An effort to encourage other states to control the export of nuclear material and equipment.

Hence, Iran is clear that it is not willing to give up its rights as an NNWS to develop its nuclear capabilities for peaceful purposes, including enrichment related activities. It also tries to revitalise the missing balance between disarmament obligations and the transfer of nuclear technology for peaceful purposes as envisaged by the NPT.

Following the Iranian proposal, China, France, Germany, the Russian Federation, the United Kingdom and the United States of America as well as the Secretary General and High Representative of the European Union communicated an additional proposal to Iran, building on its proposal on 18 June 2008 further elements to be considered in the negotiations, besides other incentives regarding the cooperation in different fields in order to reach a comprehensive agreement.[104] In this communication, they expressed their readiness "to recognise Iran's right to develop research, production and use of nuclear energy for peaceful purposes in conformity with its NPT obligations" and "to treat Iran's nuclear programme in the same manner as that of any NNWS party to the NPT once international confidence in the exclusively peaceful nature of Iran's nuclear programme is restored." They have also reiterated their position, disregarding the right of Iran to get engaged in enrichment related activities or to construct reactors run by heavy water. They are offering Iran legally binding nuclear fuel supply guarantees and support for the construction of LWR.

Although the attempts by the aforementioned powers to reach an agreement with Iran are political in nature, they target the legal foundation of the NPT. The proposals recognise the rights of Iran as an NNWS to develop and use nuclear energy for peaceful purposes, but at the same time, their position aims at targeting these rights. They want to ensure that certain activities and technologies, even if they are peaceful in nature, under IAEA safeguards and permitted by the NPT, should remain in the hands of the few and should not be transferred to NNWS. This amounts to factual amendments to the NPT, though not legally binding, but will constitute a precedence that will be imposed in reality on all NNWS in contradiction to the NPT and IAEA Statute provisions.

---

[104] IAEA, INFCIRC/750, 'Communication dated 25 June 2008 received from the Resident Representative of the United Kingdom to the Agency concerning a letter and offer of 12 June 2008 delivered to the Islamic Republic of Iran,' 1 July 2008.

6. *Conclusions*

The hostile attitude of the US towards Iran shows how an industrialised country can create barriers against the efforts of a nation seeking development and industrialisation. This seems to be a new form of imperialism: Technological Imperialism, "the policy and practice of seeking to dominate the economic or political affairs of underdeveloped areas or weaker countries."[105] This exemplifies how great powers seek to impose their preferred political patterns on other nations. Iran has always been an oil-producing country and there is nothing new in this fact. At the time when Western countries encouraged Iran to start a nuclear power plant, Iran was producing even more oil than it does today. Thus, resorting to the argument that Iran does not need other sources of energy or declarations like those uttered by the former US Secretary of State, Warren Christopher, that "they do not have shortage of energy," seem to be hypocritical to say the least. As mentioned, Iran had the same position in the 1970s, but the US administration never rejected such deals and the US was willing to sell military technology and nuclear plants to Iran.[106]

Contrary to claims made by Western newspapers, Iran as a party to the NPT has accepted IAEA safeguards, and all its activities involving nuclear materials have been open to reputable international inspection agencies. In fact, IAEA officials have visited the existing sites since 1992 and reported that "the activities at the sites were consistent with peaceful uses." During February 1992, at least six sites were selected and inspected by the IAEA. "Three of the sites visited had not been seen before by the IAEA inspectors."[107]

Meaningful enforcement of a non-proliferation inspection effort would also require better international cooperation in reporting the export of dual-use technology, and arrangements that allowed the IAEA to directly interrogate countries as to the purpose of such exports, and to then inspect the supplier's facilities. Only a few countries genuinely cooperate in pooling sensitive export data, and none allow foreign inspection of possible suppliers. Most countries also carefully compartmentalise their efforts to control nuclear proliferation so that only a few officials are aware of the full details.[108]

---

[105] Roshandel, Jalil, 'Iran, Nuclear Technology and International Security,' The Iranian Journal of International Affairs, Volume 8, Issue 1, 1996, p. 156.

[106] Ibid.

[107] Ibid, pp. 157–158.

[108] Ibid, pp. 24–25.

So far, US efforts have delayed the progress of the Iranian nuclear program for peaceful uses. However, they could not totally obstruct it. Iran was able to establish an advanced nuclear program. On the other hand, Iran did not act in transparency with the IAEA for so many years; it has concealed information about its program and how far it was in developing its nuclear infrastructure. Nevertheless, the expansive and intensive inspections that that IAEA has conducted in Iran did not reveal the development of a military nuclear program. Although Iran has not been in full compliance with its Safeguards Agreement with the IAEA, it has cooperated extensively with the Agency during its inspections from the year 2002 until now.

It might be worth asking why Iran has concealed all this information from the Agency and the international community so long as it is developing its capabilities for peaceful purposes. One likely answer is the international monopoly over nuclear technology for peaceful purposes. It might be appropriate here to question how many NNWS were able to develop their nuclear capabilities for peaceful purposes since the NPT entered into force. Very few or almost none! The countries that joined the NPT with developed nuclear infrastructure for peaceful purposes continued to dominate the market. If any state outside this club exerted any effort to reach the same developed stage, it was fought and obstructed from doing so. For example, Iran has developed a programme to produce heavy water and it has declared to the Agency that the surplus of this program can be sold to other States. Norway does that, so why should Iran be banned? If the NPT States desire this system to remain strong and effective, they should implement all of its provisions; not only the disarmament procedures on NNWS. They should not hamper the inalienable right of NNWS to develop their nuclear capabilities for peaceful purposes. If this took place, it would encourage NNWS to declare all their activities as long as they were not breaking the rules of the NPT and not aiming at developing nuclear military uses. Otherwise, this constitutes a failure in the NPT system. It came into being to encourage the peaceful uses of nuclear technology around the world and to prevent the development of nuclear weapons. If the system is not good enough to provide NNWS with enough assurances to enable them to develop nuclear capabilities for peaceful purposes, even if these assurances prevent the development of nuclear weapons, it can easily be concluded that this system failed to reach its goals.

The current rules are inadequate to ensure Iranian compliance with its obligations under its Safeguards Agreement, as they rely only on the cooperation of Iran with the Agency in providing sufficient information about nuclear material and equipment in the country. Therefore, there should be international legal rules to govern the cooperation of Supplier States with the IAEA, not only the LSG rules, in providing the Agency with sufficient information about their transfers to NNWS. This will assist the Agency to fulfil its goals. It will bridge the gap created by any limited cooperation from Iran, or any other NNWS and it will not bring the IAEA into confrontation with the State involved.

Therefore, there should be international legal rules (above and beyond the LSG rules) to govern the cooperation of Supplier States with the IAEA with regard to providing the Agency with sufficient information about their transfers to NNWS.

CHAPTER SIX

CASE III: ISRAEL

It has always been our intention to develop the nuclear potential. We now have that potential
Israeli President Ephraim Katzir, 1974

Israel, although it denies it, has it. Honoré M. Catudal

In 1969, while in a state visit to Washington, President Nixon had asked the visiting Israeli Prime Minister Golda Meir if Israel had any "dangerous toys," by that he meant an atom bomb, and asked if Israel has one? To this, the Israeli leader replied, 'We do.' Nixon seemed impressed and had cautioned her to 'be careful.' On her return, she narrated the incident to the Israeli Cabinet and remarked that she was 'just lucky that he didn't ask how many bombs.'
James Adam, The Unnatural Alliance

The Israeli nuclear program is of special interest to this study because its legal position totally differs from the other previous case studies Israel is not a member of the NPT, and a very limited number of its nuclear facilities fall under IAEA safeguards. However, the Israeli nuclear program and armament was subject to several studies by the SG of the UN and the United Nations Institute for Disarmament Research (UNIDIR). It was also the subject of one SC resolution and several GA resolutions. Furthermore, Israel's formal position follows a policy of nuclear ambiguity by declaring, "It will not be the first to introduce nuclear weapons into the region."[1] It also supports the establishment of an NWFZ in the Middle East.

In this chapter, the Israeli nuclear program shall be discussed as follows:

1. Development of the Israeli nuclear program for peaceful and military purpose.

[1] GA, A/40/520, Report of UNIDIR, 'Israeli Nuclear Armament,' Distribution General, 9 August 1985, p. 4.

II. Legal aspects surrounding the program and legal effects of UN resolutions concerning Israel nuclear armament.

1. *Development of the Israeli Nuclear Programme for Peaceful and Military Purposes*

Israeli efforts to acquire the technology, personnel and materials to produce nuclear weapons date back almost as far as the birth of the State in May 1948.[2] Israel has been actively engaged in various aspects of nuclear research and has reportedly developed its own sources of uranium and has acquired expertise of various processes that make up the nuclear fuel cycle.[3]

Today, Israel has four major university institutions that deal with fundamental nuclear research and education: the Weizmann Institute of Science at Rehvoth, the Rach Institute of Physics at the Hebrew University of Jerusalem, the Israel Institute of Technology-Technion at Haifa and the Ben-Gurion University of the Negev at Beer-Sheba. The Israeli Government, through Israel Atomic Energy Commission (IAEC), controls the Nahal-Soreq Nuclear Research Centre and the Negev Nuclear Research Centre and their reactors.[4]

In the 1950s, the US unintentionally aided Israeli efforts when President Eisenhower introduced his "Atoms for Peace" programme in 1955. Under this programme, Israelis received technical training in the US and, most important, Washington agreed to construct a 5-MW reactor for Israel near Nahal Soreq.[5]

[2] Barnaby, Frank, 'Capping Israel's Nuclear Volcano,' in Efraim Karsh (ed.), 'Between War and Peace: Dilemmas of Israeli Security,' 1996, p. 95.

[3] GA, A/40/520, 1985, p. 4. The Report also stated, "35. The natural resources of Israel are relatively limited. There are no uranium ore deposits, but the phosphate reserves in the Negev Desert contain uranium at concentrations of 100–170 parts per million. The proven and estimated reserves of uranium in the phosphates of Israel are estimated, respectively, to be 40,000 and 200,000 tons. 36. Extensive research on methods for extracting uranium from the Negev phosphates has been conducted by the Israeli Nuclear Research Centre at Dimona. Their scientists claim to have successfully tested a new technique involving the use of a novel ion exchange resin, which is applied to the separation of uranium from phosphoric acid. A pilot plant has been in operation since the beginning of 1984 at the Rotem Fertiliser Ltd. plant at Arad."

[4] Ibid, pp. 9–10.

[5] Catudal, Jr., Honore M., 'Israel's Nuclear Weaponry—A New Arms Race in the Middle East,' 1991, p. 115.

However, it was France that provided the crucial assistance leading to Israel's acquisition of the bomb. The cooperation between the two countries started in 1949, when some Israeli scientists were permitted to join Saclay, the newly established French national atomic research centre near Versailles, and participate in the construction of Saclay's small experimental reactor, known as EL 102, which was powered by natural uranium and moderated by heavy water. Building a similar reactor in Israel was eminently feasible. Uranium was indigenous to Israel, and there was some heavy water available locally in Israel; more heavy water, if needed, as seemed likely, could be supplied by the French or illicitly purchased from Norway or the United States, then the world's largest producers. They were also the only foreigners allowed access throughout the secret French nuclear complex at Marcoule.[6]

After 1955 and before the 1956 Suez War with Egypt, the French secretly agreed to supply Israel with a sizable plutonium reactor to build at Dimona in the desert. There was an obvious trade-off: Ben Gurion agreed to withdraw his troops from Sinai and accept the UN peacekeeping role in return for French help in building a nuclear reactor and chemical reprocessing plant. When De Gaulle came to power in 1958, he began to slow French cooperation with Tel Aviv. He finally ended it altogether in later years, but not before France had allegedly shared nuclear test data in 1960.[7]

By the spring of 1957, it was decided to locate the reactor at Dimona near the ancient city of Beersheba in the Negev. A French chemical firm was selected to build the Israeli reprocessing facility—underground, and with plutonium production capacity enough for four nuclear bombs a year.[8]

In early 1958, U2 American spy planes reported that subterranean digging has started at Dimona. Since that date, the U2 spotted activities at the site. They have monitored digging for a second site that was meant for the chemical reprocessing plant. In fact, by end of 1959, U2 operators had no doubt that Israel was going for a nuclear bomb.[9]

---

[6] Hersh, Seymour M., 'The Samson Option: Israel's Nuclear Arsenal and American Foreign Policy,' 1991, pp. 26–30.

[7] Catudal, Jr., Honore M., 'Israel's Nuclear Weaponry—A New Arms Race in the Middle East,' 1991, p. 115.

[8] Hersh, Seymour M., 'The Samson Option: Israel's Nuclear Arsenal and American Foreign Policy,' 1991, p. 31.

[9] Ibid, pp. 45–55.

By the early 60s, Israel was trying to acquire uranium to supply its reactor. The FBI was able to determine that 1963 was the year in which diversions of enriched uranium to Israel might have begun to take place from a company near Pittsburgh, Pennsylvania. The Nuclear Material and Equipment Corporation (NUMEC) in Apollo, Pennsylvania, formerly headed by Zalman Shapiro, had a contract with the American Government to turn bomb-grade uranium into fuel for submarines. "During AEC inspections of 1964 and 1965, the AEC discovered that large quantities of uranium given to NUMEC for fuel conversion had never been returned to the government. Weapons-grade uranium had somehow disappeared. Subsequent investigations by US intelligence agencies uncovered evidence that from 1957 to 1967, NUMEC President Shapiro had conspired with Israel to pilfer some 200 pounds of enriched uranium from his own company." This was confirmed during 1964, when the then Director of the CIA, Richard Helms, received an alarming report. CIA agents had uncovered the fact that Israel had managed to obtain an unspecified quantity of enriched uranium. As one source noted: "Although the Agency was unable to analyse any of the U-235, it was perfectly clear it had been obtained illegally—most probably from the US which had the largest and most sophisticated enrichment operation in the world."[10]

The first fuel charge of Dimona research reactor, in 1963, contained 24 tons of uranium. Of this load, ten tons came from "Israeli domestic production", 4 came from France and ten came from South Africa which shipped uranium to Israel as early as 1957. These shipments kept the Israeli nuclear weapon production going.[11]

Before June 1967, the American Embassy in Tel Aviv reported that Israel had completed its basic weapons design and was capable of manufacturing warheads for deployment on missiles. Israel also may have had a crudely manufactured bomb or two ready to go, but—as the embassy could not know—no decision had been made by Prime Minister Eshkol to begin mass production.[12]

The exact date when the Israelis began to construct a deliverable nuclear weapon is not known to the West, but informed speculation

[10] Ibid, pp. 36–38.

[11] Beit-Hallahmi, Benjami, 'The Israeli Connection—Whom Israel Arms and Why?' 1987. p. 133.

[12] Hersh, Seymour M., 'The Samson Option: Israel's Nuclear Arsenal and American Foreign Policy,' 1991, p. 162.

has it that Israel had nuclear weapons and the ability to deliver them at the beginning of the Six Day War in 1967. In fact, some scholars assert that one major reason the Arabs fought that war was to pre-empt an Israeli nuclear capability. However, the decision to build up Israel's nuclear weapons probably came after the 1967 war. Indeed, throughout the early 70s, the development of Israeli nuclear weapons became increasingly apparent. This was true despite the official Israeli position that "Israel is not a nuclear power and will not be the first to introduce nuclear weapons in the Middle East."[13] Since then, it has been commonly assumed that Israel has been a NWS. The Israeli nuclear program, however, has remained obscure—shrouded in secrecy, officially unacknowledged, and insulated from domestic Israeli politics.[14]

However, "there is no doubt that Israel has the technical capability to manufacture nuclear weapons and possesses the means of delivery of such weapons to targets in the area. To recapitulate: Israel has an unsafeguarded reactor capable of producing considerable amounts of plutonium and has some means of separating plutonium from irradiated uranium fuel. It has the technological skill and expertise as well as the technical infrastructure required to manufacture nuclear weapons. Since the greater part of the Israeli nuclear programme is not under safeguards, few technical details about its actual nuclear activities are known. However, since 1964, when Dimona went into operation, Israel could have produced sufficient weapons-grade plutonium for a significant number of explosive devices."[15]

It is widely assumed that the scientific nuclear programme of Israel includes some research into plutonium extraction. IAEA and SIPRI have reported that a pilot facility for reprocessing spent fuel exists in Israel. According to SIPRI, the principal equipment for the facility was supplied by a French firm. There is no official confirmation regarding the capacity of this facility, although there are reports claiming its capability of handling up to 3,400 kg of irradiated fuel a year from which it can extract from 4 to 5 kg of plutonium. It is also possible to separate small quantities of plutonium in radiochemistry laboratories (so-called "hot cells"). Many advanced universities, in fact, possess radiochemistry laboratories where such work may be carried out on a small scale. For purposes of

[13] Ibid, pp. 116–162.
[14] Cohen, Avner, 'Israel and the Bomb,' 1998, p. 1.
[15] GA, A/40/520, Report of UNIDIR, 'Israeli Nuclear Armament,' Distribution General, 9 August 1985, p. 4.

manufacturing nuclear weapons, it is technically desirable to achieve as high a fissile content of the plutonium (i.e. more than 90 % of the isotope 239) as possible.[16]

The 1973 war brought Israel's growing nuclear arsenal into the bright glare of international publicity for the first time, when the Israelis reportedly prepared them for use in a last desperate move. This war experience suggests that the Israelis will continue to rely on nuclear weapons as a deterrent, even as this policy is now becoming more overt. Israel apparently plans to use the bomb only in response to actual aggression threatening the destruction of Israel—not as a first strike tool to pre-empt anticipated aggression.[17]

In early 1961, two American physicists, Isidor Isaac Rabi of Columbia University and Eugene Wigner of Princeton, visited the yet incomplete reactor at Dimona. Neither reported seeing evidence of a weapons facility. Later, Ben Gurion, the former Israeli Prime Minister, refused the proposal of the US President Kennedy to allow the IAEA to inspect Dimona. However, they agreed that the US would send a specially assembled American inspection team into Dimona. The team would have to schedule its visits well in advance and with full acquiescence of Israel. There would be no spot checks permitted.[18]

The Israeli scheme, based on plans supplied by the French, was simple: a false control room was constructed at Dimona, complete with false control panels and computer-driven measuring devices that seemed to be gauging the thermal output of a twenty-four-megawatt reactor (as Israel claimed Dimona to be) in full operation. There were extensive practice sessions in a fake control room, as Israeli technicians sought to avoid any slips when the Americans arrived. The goal was to convince inspectors of the inexistence or even the possibility of existence of a chemical reprocessing plant. One big apprehension was that the Americans would seek to inspect the reactor core physically, and presumably discover that Dimona was utilising large amounts of heavy water—much of it illicitly obtained from France and Norway—and obviously operating the reactor at far greater output than the acknowledged twenty-four megawatts. It

[16] GA, A/36/431, 'Israel Nuclear Armament,' Report of the SG of the UN, 18 September 1981, p. 20.

[17] Catudal, Jr., Honore M., 'Israel's Nuclear Weaponry—A New Arms Race in the Middle East,' 1991, pp. 116–117.

[18] Hersh, Seymour M., 'The Samson Option: Israel's Nuclear Arsenal and American Foreign Policy,' 1991, pp. 92–102.

was agreed that the inspection team would not be permitted to enter the core "for safety reasons." The American team, following a pattern that would be repeated until the inspections came to an end in 1969, spent a day at Dimona, climbing through various excavations—many facilities had yet to be constructed—but finding nothing. They did not question the fact that the reactor core was off-limits and gave no sign that they were in any way suspicious of the control room. The Israelis even stationed a few engineers in a concealed area in the control room to monitor the machinery and make sure that nothing untoward took place.[19]

On 7 March 1969, in an "Action Memorandum," Assistant Secretary of State Joseph Sisco outlined the background of American visits to Dimona, mentioning that since May 1961, the US had conducted "seven inspections of the [Dimona] facility"—the memo used the words "inspection" and "visit" interchangeably—the last one taking place in June 1968. "Our understanding with the Government of Israel is that visits will be conducted without publicity, but that we will be free to convey the results to other governments of our choice." The memo noted that the Israelis had always insisted on maintaining visits at least a year apart, "citing domestic political difficulties." The Israeli ground rules—"one-day visits on the basis of one-year periodicity"—had been minimally sufficient to give the Americans reasonable confidence that Israel was not engaged in weapons-related activity at this site.[20]

Notwithstanding the deficiencies of the American visits to Dimona, there were growing suspicions that Israel might have other nuclear weapon-related facilities. Since around 1966, the American intelligence community was of the opinion that real weapons work might be occurring not in Dimona but "somewhere else in Israel." Sisco's memorandum expressed this suspicion: "I would stress that while our inspections in Dimona can give us information about activities at that site, they cannot exclude the possibility (which we in fact believe to be a likelihood) that the Israelis are engaged in nuclear weapons R&D somewhere else in Israel. Nevertheless, since Dimona is the only installation in Israel to our knowledge that can produce fissionable material in sufficient quantities for a weapon programme, we consider it important to check periodically as to whether the operations at this facility are devoted exclusively to peaceful purposes."[21]

---

19 Ibid, pp. 110–111.
20 Cohen, Avner, 'Israel and the Bomb,' 1998, p. 329.
21 Ibid, pp. 329–330.

In either 1967 or 1968, the CIA monitored uranium traces around the Dimona complex and was able to determine that a nuclear weapons program at the plant was in advanced stages. "It is a safe assumption," writes Stephen Green, "that Israel had nuclear weapons and the ability to deliver them at the beginning of the Six Day War in 1967." The "ingredients were there, hence the bomb was ready." In addition, Israel continues to hint unofficially at its possession of nuclear arms—which are now said to include the H-bomb—while officially denying that such arms exist. But this strategy of "deliberate ambiguity," as it is known in strategic literature, became less tenable as a result of detailed revelations, in 1986, by Mordechai Vanunu, a former nuclear technician at the Israeli highly classified nuclear facility in the Negev Desert. He outlined in detail to the Sunday times how Israel produced plutonium for nuclear weapons and he unambiguously confirmed with pictures and documentation the existence of an extensive Israeli nuclear weapons programme.[22]

As described by Vanunu (and confirmed in later interviews with Israeli officials), Dimona includes the reactor and at least eight other buildings, or Machons, the most important of which is a chemical reprocessing plant. Each building apparently is self-contained. Machon 1 is a large silver-dome reactor, sixty feet in diameter and is clearly visible from the nearby highway. Uranium fuel rods remain for three months in the reactor, which is cooled and moderated by heavy water. Heavy water is cooled by ordinary water flowing through a heat exchanger, creating steam, which in a nuclear power plant would drive a turbine and create electricity; instead, the steam in Machon 1 is vented into the atmosphere, creating a radioactive cloud. Machon 2 is a chemical reprocessing plant where plutonium, a by-product of the fission process in the reactor, is extracted by chemical means from spent uranium rods. The end result of the chemical processing, according to Vanunu, is a weekly average of nine "buttons" of pure plutonium whose combined weight is 1.2 kilograms. Machon 3 converts lithium 6 into solid form for insertion into a nuclear warhead and also processes natural uranium for the reactor. Machon 4 contains a waste treatment plant for radioactive residue from the chemical reprocessing plant in Machon 2. Machon 5 coats uranium rods (shipped from Machon 3) with aluminium to be consumed in the reactor. The rods, once stacked in the core of the reactor, provide the fuel needed to sustain a chain reaction—and capture of weapons-grade

[22] Ibid, p. 117.

isotopes of plutonium. Machon 6 provides basic services and power for Dimona. Machon 8 contains a laboratory for testing samples and experimenting on new manufacturing processes; it is also the site of Special Unit 840, where Israeli scientists have developed a gas centrifuge method of enriching uranium for weapons use. There is also a laser-isotope reprocessing facility for enriching uranium in Machon 9. Depleted uranium is chemically isolated in Machon 10 for eventual shipment to the Israeli Defence Force or sale to arms manufacturers in Europe and elsewhere for use in bullets, armour plating, and artillery and bomb shells. The shells, buttressed by heavy uranium, which is much denser than lead, can easily penetrate thick armour plating and are a staple in modern arsenals. (There was no Machon 7 when Vanunu worked at Dimona.) Weapon grade plutonium is stored in metal form inside sealed glove boxes ready for insertion into a nuclear warhead. Other chemicals used in the Israeli nuclear arsenal, including lithium compounds and beryllium, are also fabricated there. However, the final stage of warhead production takes place at a defence plant north of Haifa operated by Rafael, the top-secret Israeli research and manufacturing agency that is responsible for Israel's most sensitive weaponry.[23]

Vanunu estimated that the Israeli nuclear arsenal might contain as many as 200 nuclear weapons. Knowledgeable US officials do not dispute the accuracy of the bulk of the testimony of Vanunu on this matter; however, they conservatively put the number somewhere between fifty and sixty. But either way it is a regional monopoly.[24]

In the aftermath of the revelations of Vanunu, Israeli leaders have continued to reiterate their long-standing position that Israel does not possess nuclear weapons and "will not be the first country to introduce nuclear weapons into the Middle East." But on a secondary official level, Israeli nuclear threats are becoming more overt. For instance, shortly after the Sunday Times ran the Vanunu story, Gideon Raphael, the former Director General of Israel Foreign Ministry and the Israeli Ambassador to the UN in the late 1960s, made a comparable reference to the Israeli nuclear arsenal. In an open piece in the Washington Post, he declared that the Syrian chemical warfare capabilities "could mean that the next war between Syria and Israel will degenerate into a contest between chemical and radiation weapons—with global implications." Amos Rubin, an

[23] Ibid, pp. 201–203.

[24] Hersh, Seymour M., 'The Samson Option: Israel's Nuclear Arsenal and American Foreign Policy,' 1991, p. 118.

economic adviser to the Prime Minister, stated in an interview to the Christian Science Monitor that "if left to its own [Israel] will have no choice but to fall on a riskier defence which will endanger itself and the world at large ... To enable Israel to abstain from dependence on nuclear arms calls for $ 2 to $ 3 billion US aid."[25]

Until recently, the Israeli policy remains that of "bomb in the basement."[26]

In 1979, Israel and South Africa were believed to have been preparing together to test a bomb in the Kalahari Desert. Since the 60s, both countries had been in need of tactical nuclear weaponry. This led to the development of a nuclear shell, fired from a 155 mm howitzer or from a naval gun, which was tested in 1979. This shell contained a low-yield, two-kiloton nuclear device—an ideal nuclear weapon for the two countries.[27]

Just before dawn on the stormy morning of 22 September 1979 clouds over the South Indian Ocean suddenly broke and an American satellite was able to record two distinctive bright flashes of light within a fraction of a second—probable evidence of a nuclear explosion. The nuclear detection satellite, known as VELA, had seen similar flashes of light on forty-one previous occasions, and in each case it was subsequently determined that a nuclear explosion had taken place. There were a few intelligence officials and non-proliferation experts in the Carter administration who immediately concluded that Israel and South Africa had finally conducted a nuclear test, a test that they had tried, and failed, to accomplish two years before. They were right. According to former Israeli government officials, the warhead tested that Saturday morning was a low-yield nuclear artillery shell that had been standardised for use by the Israeli Defence Force. They also said that the event captured by the VELA satellite was not the first but the third test of a nuclear device over the Indian Ocean.[28]

Another Israeli, who also had direct access to Defence Ministry information about the test in South Africa, said that Weizman signed an agreement before the 1979 tests calling for the sale to South Africa of technology and equipment needed for the manufacture of low-yield 175 mm and

[25] Ibid, p. 122.
[26] Ibid.
[27] Beit-Hallahmi, Benjami, 'The Israeli Connection—Whom Israel Arms and Why?' 1987, pp. 133–136.
[28] Hersh, Seymour M., 'The Samson Option: Israel's Nuclear Arsenal and American Foreign Policy,' 1991, pp. 271–272.

203 mm nuclear artillery shells. Weizman's order triggered an internal dispute with senior nuclear officials, the Israeli recalled, who protested against the government's decision to sell the information, considered by officials running Dimona to be "the best stuff we got."[29]

The first affiliation of South Africa and Israel to the explosion of 22 September 1979, came on 21 February 1980, when CBS-TV news reported that the 1979 flash was indeed an Israeli-South African nuclear test. This was confirmed by the CIA, and in a special programme produced and broadcasted by the Israeli State Television on 21 December 1980. The programme dealt in detail with Israeli- South African nuclear cooperation. It reported that the 1979 flash resulted from a test of the newly developed naval nuclear shell, part of the joint programme. There were no reactions on the Israeli media following the broadcast of the TV special.[30]

Despite the unwillingness of some politicians to acknowledge the test, nuclear collaboration between Tel Aviv and Pretoria continued: a second test was reported on 16 December 1980 in the same area, with another CIA sighting of South African ships nearby. "A British authority on nuclear weapons, Dr. David Baker, said that the weapon fired in this test was probably a 155 mm nuclear shell fired from a special howitzer which the Israeli had helped the South Africans acquire."[31]

In an unprecedented and unexpected statement, the Israeli Prime Minister Ehud Olmert listed Israel among NWS.[32] His remarks raised reactions by commentators as well as Governments, who considered this to be the first statement by a Prime Minister of Israel to admit publicly that his country is a nuclear State. The Coordination Bureau of the Non-Aligned Movement in New York (CoB of NAM) issued a statement as a

---

[29] Ibid, p. 276.

[30] Catudal, Jr., Honore M., 'Israel's Nuclear Weaponry—A New Arms Race in the Middle East,' 1991, pp. 84–85.

[31] Ibid, p. 85.

[32] DPA German Press Agency, 'Olmert appears to admit Israel possesses nuclear weapons,' it stated, 'Israeli Prime Minister Ehud Olmert appeared to admit Monday that Israel possesses nuclear weapons, when he included the country in a list of nuclear states. In an interview with Germany's Sat 1 television, an excerpt of which was broadcast in Israel, the Prime Minister was asked whether Israel's nuclear arsenal—which until now it has never publicly admitted to having—undermined the West's objections to a nuclear Iran. "Israel is a democracy and does not threaten anyone ... Iran explicitly, openly and publicly threatens to wipe Israel off the map," an agitated Olmert replied. "Can you say that this is the same level, when they are aspiring to have nuclear weapons, as America, France, Israel and Russia?" he asked, in remarks immediately seized on by some Israeli commentators as being an admission that the Jewish state does have nuclear arsenal,' 11 December 2006.

reaction to the one delivered by Olmert. It was circulated as an official document of both the GA and the SC.[33] In their letter, the CoB of NAM stated that is "gravely concerned by the statement" and expressed "its great concern over the acquisition of nuclear capability by Israel, which poses a serious and continuing threat to the security of neighbouring and other States, and condemned Israel on its action and the statement in that regard, and for continuing to develop and stockpile nuclear arsenals." It has also "demanded that Israel, the only one in the region that has not joined the NPT nor declared its intention to do so, to renounce possession of nuclear weapons, to accede to the NPT without delay, to place promptly all its nuclear facilities under IAEA full-scope safeguards according to SCR 487 (1981), and to conduct its nuclear-related activities in conformity with the non-proliferation regime."[34]

2. *Legal Aspects Surrounding the Israeli Nuclear Programme and Legal Effects of* UN *Resolutions Concerning Israeli Armaments*

Israel, according to the previous review, possesses nuclear weapons despite the fact that it has kept its official position not to declare its possession. The announcement by Olmert leaves no doubts that Israel is a nuclear power.

2.1. *Israel and the* NPT

Although the official position of Israel is supportive of nuclear non-proliferation, it is still not a member of the NPT. Even in the case of considering the NPT as part of customary international law,[35] the fact remains that Israel and several other States are not members to the treaty. Consequently, they do not abide by its rules. Israel has one nuclear facility, in Nehal Soreq that was under the IAEA safeguards system before the entry into force of the NPT (INFCIRC/66).

---

[33] GA, SC, A/61/735–S/2007/73, 'Letter dated 8 February 2007 from the Permanent Representative of Cuba to the United Nations addressed to the Secretary-General,' 9 February 2007.

[34] Ibid.

[35] Statement of Raouf, Tarik, Section Head, Verification and Security Policy Coordination, Office of External Relations and Policy Coordination at the IAEA, in an interview concluded at IAEA Headquarters, 2004.

Since 1968, Israel has specified reservations related to the NPT. These reservations can be summarised as follows:[36]

The NPT entailed no commitment by nuclear powers that they would not use their nuclear weapons against those who did not possess them. Although SC resolution 255 and the identical declarations of the three main nuclear powers (US, USSR and UK) had a moral authority which should not be ignored, they were not considered by Israel as compensatory measures for several reasons; First, there was lack of universality—both with regard to the adherence to the Treaty and with regard to the proposed security assurances. Second, the resolution and the identical declarations did not constitute any binding commitment. Third, although the nuclear powers were ready to pledge all the smaller nations a guarantee against nuclear war, Israel was concerned mostly with conventional war. On these grounds, further negotiations were needed before it signed this important document. Therefore, Israel maintained that there would be greater confidence in its value if NNWS could be certain that those who subscribed to the document accepted without reservations the principle set forth in the last paragraph of the preamble of the NPT, with respect to the obligation to refrain from the threat of use of force. Fourth, from the perspective of conditions prevailing in the Middle East and their bearing on and relationship to the NPT, Israel considers that implementing the NPT requires the existence of conditions for peace. However, such conditions did not exist in the region as the Middle East is an area of diversities, where rivalries and conflicts trigger constant instability. Fifth, Israel believes that, in the Middle East, the implication of a conventional arms race could not be ignored. This constitutes an immediate threat to Israel more than nuclear weapons. Hence, Israel maintains that as long as the resort to war is not ruled out, little faith can be placed either in unilateral declarations amounting to no more than self-denial ordinances in the nuclear sphere, or in strict and unbroken compliance with such pledges. In fact, Israel considers it important to establish a system of reassurances and mutually binding obligations among all States to abstain from introducing nuclear weapons into the region. This can be achieved through regional arrangements between all States in the area. In addition, Israel maintains that declarations and reservations made by several Arab States concerning the inapplicability of their obligations under arms control and disarmament agreements vis-à-vis Israel are incompatible

[36] Marom, Ran, 'Israel's Position on Non-Proliferation,' June 1996, p. 27.

with the aims and spirit of the treaties in question and therefore, constitute an obstacle to the establishment of an effective non-proliferation regime.[37]

There was no confidence that some Arab States which adhered to the NPT would indeed abide by their obligations under the Treaty. Examples of such states are Libya and Iraq. Another issue of concern was the Pakistani relationship with some Arab States in the nuclear field.[38]

In line with this view, Libya and Iraq were indeed seeking nuclear weapons and Pakistani nuclear scientists sold some technology and equipment to Libya.[39] However, a major driving force behind these activities is the position undertaken by Israel to maintain its nuclear monopoly in the Middle East.

Moreover, all States in the region are members of the NPT and their nuclear facilities are subject to IAEA/NPT safeguards. Hence, if any of these States pursue nuclear activity for military purposes, it will constitute a breach of international legal obligations. Besides, the international community takes very strict measures against any breach of NPT obligations in order to ensure that such a State complies with its obligations. In fact, the developments in the nuclear field in the two States previously mentioned, Libya and Iraq, represent a good example in this respect.

Another problem is the Israeli claim that Arab States manipulate the IAEA policy-making organs twisting them into instruments of political warfare and illegal action against Israel.[40]

This view is misleading as there were no illegal actions undertaken by the Arab States against Israel via the IAEA policy-making organs. They co-operate with the Agency to help it fulfil its job. There are political differences between Israel and the Arab States in the region, some which are evident in the Arab States' positions expressed within international agencies, among which is the IAEA. That does not entail the existence of political warfare against Israel. Otherwise, the Arab States would have the right to judge that Israel uses political warfare against them too. In simple terms, political differences between States affect their positions within the work of international organisations and cannot be described as political warfare or invoking illegal actions.

---

[37] Ibid, pp. 27–31.

[38] Marom, Ran, 'Israel's Position on Non-Proliferation,' June 1996, pp. 33–38.

[39] Weiss, Leonard, 'Pakistan: it's déjà vu all over again,' Bulletin of Atomic Scientists, Volume 60, No. 3, May/June 2004, p. 52.

[40] Marom, Ran, 'Israel's Position on Non-Proliferation,' June 1996, p. 39.

Israel's position is not fully comprehensible. Israel insists on not joining the NPT individually, as the rest of the States in the Middle East have done. A question suggests itself here: what are the reasons behind the Israeli hesitance to join the treaty? If Israel decides to join the NPT, and to put its nuclear facilities under full scope safeguards, this will curb the nuclear proliferation threats in the Middle East and against Israel itself. Furthermore, regarding other weapons of mass destruction, like chemical and biological weapons, literature reveals that several States in the region are developing them, including Israel.[41] Therefore, dismantling such weapons requires negotiations between States in the region. However, if the only NWS in the region (Israel) decides to give up its nuclear arsenals, this will constitute a positive step towards establishing an NWFZ in the Middle East and thence establishing a zone free from weapons of mass destruction could be feasible.

Israeli politicians, who are opposed to joining the NPT, consider that, in the absence of reliable arrangements for preventing armed conflict, nuclear deterrence is essential for the survival of the nation. Publicly, they claim that the NPT would be used to increase hostile pressures against Israel, and that the safeguards under the Treaty are not reliable enough to protect Israeli interests. They have stated, however, that they are prepared to negotiate, with all parties concerned, the establishment of an NWFZ in the Middle East but they demand strict regional assurances of compliance, even though no other state in the region is likely to be able to develop a full-fledged nuclear weapon capability in less than a decade or even more. The Arab countries, on their part, insist that all States of the envisaged NWFZ should first join the NPT, or at least declare their determination not to manufacture or otherwise acquire nuclear weapons. Recent reports that Iran, Egypt, Libya and Syria possess chemical weapons and are acquiring missiles capable of delivering these weapons are mirrored by the fact that the Israeli Government has ignored UN resolutions calling upon Israel to submit all its nuclear installations to international safeguards. However, given geopolitical circumstances in the Middle East, nuclear deterrence can hardly guarantee Israeli security over the long term. The emergence of a NWS in the region, in response

[41] Hersh, Seymour M., 'The Samson Option: Israel's Nuclear Arsenal and American Foreign Policy,' 1991, he stated, 'Israeli scientists also had been tracked to a nearby French chemical and biological weapons (CBW) testing area in the Sahara.' This took place in the early years of nuclear cooperation between Israel and France, p. 58.

to Israeli nuclear armament, would nullify the advantages of Israel's monopoly and neutralise its deterrent. Indeed, owing to its small size and the density of its population, Israel would have more to lose from a nuclear exchange than any neighbouring state.[42]

2.2. *UN Resolutions Concerning Israel Nuclear Armament*

The nuclear programme of Israel has been the subject of several UN resolutions, one from the SC and the rest from the GA. In this section, an assessment of the legal effect of resolutions adopted by the UN organs regarding the Israeli nuclear programme will be presented, while bearing in consideration that all States, members of the UN, have a duty to fulfil, in good faith, their obligations under the UN Charter as demonstrated in the Declaration of Principles of International Law concerning Friendly Relations and Cooperation adopted by the GA on 24th October 1970 on the occasion of the 25th anniversary of the UN. The declaration took a progressive step forward in amplifying the Charter provisions concerning prohibition of force and in seeking to invest those provisions with substantive content.[43]

The declaration set forth the seven principles of international law as enshrined within the UN Charter, concerning the obligation to refrain from threat or use of force in contravention of the UN Charter, the obligation of States respecting peaceful settlement of international disputes, the duty of non intervention, international cooperation, sovereign equality and the duty to fulfil Charter obligations in good faith.[44]

However, strict measures have never been taken by UN bodies to urge Israel to cease from developing nuclear weapons as this is related to political issues and different perspectives of powerful members of the UN. Professor Schachter highlighted this point when he stated, "The metropolis of peace and security law, like human rights, has its suburbs. One of the most important is the regulation and prohibition of armaments, especially weapons of mass destruction. The legal pattern in this area is predominantly made up of negotiated agreements, on a global or regional basis. It also embraces inspection to ascertain compliance with legal obli-

[42] Goldblat, Josef, 'Twenty Years of the NPT implementation and prospects,' 1990, p. 13.
[43] Raghavan, C.G., 'The UN Charter and the Use of Force in Interstate Relations,' R.C. Hingorani, (ed.), 1972, p. 165.
[44] Ibid, p. 165.

gations such as those contained in the NPT or SC decisions under chapter VII as, for example, applied to Iraq. Efforts to outlaw mass weapons or their use on the basis of existing principles of customary law or the Charter itself persist in UN bodies but lack support by the major powers concerned."[45]

The existing principles of customary international law and the UN Charter are sufficient to outlaw WMD. Based on the rules governing the legitimate use of force, its main principles are necessity, proportionality and that the State's action should be one of self-defence. As previously stated, all wars that have taken place between the Arab States and Israel were conventional wars and no WMD was ever used. But, the fact remains that WMD preserve the balance of power in the region in addition to the role that conventional weapons already play in this regard. Hence, if the previous principles of use of force are implemented, it can be concluded that WMD are illegal, especially if we consider it unnecessary to initiate their use and that these weapons are indiscriminatory by nature, and therefore defy the principle of proportionality.

SC Resolution 487

The Israeli nuclear programme was mentioned in SC resolution 487 of 1981, in which the SC, inter alia, called upon Israel "to adhere to the NPT and to place its nuclear facilities under full scope safeguards".

It has been maintained that the Council may adopt decisions binding the members solely within the framework of Chapters VI, VII and VIII of the Charter. Moreover, it is also claimed that decisions of the Council, in conformity with Article 25 do not make a distinction among types of decisions that the members "agreed" to accept and carry out. According to an old legal aphorism, "where the legislator does not distinguish, the interpreter must not distinguish." However, whenever the connection between the subject of a resolution and maintenance of international peace and security permits the council, juridically, to place a decision within the framework of Chapter VII (even without any direct reference to it in the text of the resolution), the Council has recognised its own competence to act for the general purpose of maintaining or restoring

[45] Cassidy, Stephen H., 'The Newest Member of the Nuclear Club: Pakistan's Drive for a Nuclear Weapons Capability and US Nuclear Non-proliferation Policy,' Hastings International Comparative Law Review, Volume 12, no. 1, 1988/1989, pp. 18–19.

peace through decisions it deemed binding, even outside the concrete procedures specified in Chapter VII, such as those provided for in Articles 41, 42, 43, 46 and 47.[46]

The source of norms adopted by the SC is, first of all, the UN Charter. Article 24 comprises a general provision outlining the functions and powers. The legal character of resolutions based on specific provisions of the Charter additionally confirmed by Article 2, paragraph 2 and 5, and Article 25, obligates Member States to carry out the decisions of the SC.[47] This was clearly expressed in the Council resolutions regarding the situation in South Africa, which are also pertinent in this respect. Although Resolution S/4300 of 1 April 1960 also established clearly (1st operative paragraph) the connection between the prevailing internal situation and the danger to international peace, the emphatic decision urgently requesting South Africa to "abandon its policy of apartheid and racial discrimination" is legally based on general powers of the SC, in conformity with paragraph 1 of Articles 24 and 25, and not based concretely on the specific powers defined in Chapter VI and VII of the Charter.[48]

If the previous interpretation is applied to SC resolution 487 of 1981, which called upon Israel to adhere to the NPT and to place its nuclear facilities under full scope safeguards, it is thereby a binding resolution and not a mere recommendation to the State of Israel.

Although this resolution, among others, does not appear to be grounded in Articles 36 or 37 of the Charter, nevertheless the SC authority has never been questioned for their adoption.[49]

The use of the wording "calls upon" is considered an order by the SC, such as the order "calls upon" to the Belgian troops to withdraw from the Congo in Resolution S/ 4387.[50]

On the other hand, the fact that an SC resolution represents a compromise does not necessarily mean that it is void of legal consequences. Article 2 of the Charter stipulates that Member States of the Organisation "shall fulfil in good faith the obligations assumed by them in accordance

[46] Castañeda, Jorge, 'Legal Effects of United Nations Resolutions,' Number 6,1969, pp. 72–76.

[47] Sonnenfeld, Renata, 'Resolutions of the UN SC,' 1988, p. 15.

[48] Castañeda, Jorge, 'Legal Effects of United Nations Resolutions,' Number 6, 1969, p. 75.

[49] Sonnenfeld, Renata, 'Resolutions of the UN SC,' 1988, p. 102.

[50] Castañeda, Jorge, 'Legal Effects of United Nations Resolutions,' Number 6, 1969, p. 74.

with the present Charter," and this provides a sufficient legal basis for the application of the provisions of Article 25. This is the interpretation adopted by the ICJ in its advisory opinion in the case of Namibia,[51] when it stated that "The decisions made by the SC in paragraphs 2 and 5 of resolution 276 (1970), as related to paragraph 3 of resolution 264 (1969) and paragraph 5 of resolution 269 (1969), were adopted in conformity with its Articles 24 and 25. The decisions are consequently binding to all States Members of the UN, which are thus under obligation to accept and carry them out. In pronouncing upon the binding nature of the SC decisions in question, the Court would recall the following passage in its Advisory Opinion of 11 April 1949 on Reparation for Injuries Suffered in the Service of the UN: "The Charter has not been content to make the Organisation created by it merely a Centre for harmonising the actions of nations in the attainment of these common ends (Article 1, Para. 4). It has equipped that centre with organs, and has given it special tasks. It has defined the position of the Members in relation to the Organisation by requiring them to give it every assistance in any action undertaken by it (Article 2, para. 5), and to accept and carry out the decisions of the SC." Thus, when the SC adopts a decision under Article 25 in accordance with the Charter, it is for Member States to comply with that decision, including those Members of the UN who are not members of the Council. To hold otherwise would be to deprive this principal organ of its essential functions and powers under the Charter. The decisions are consequently binding to all State Members of the UN, which are thus under obligation to accept and carry them out. The Court put forward the following arguments, while explaining the reasons for adopting its opinion: (1) The adoption in the name of the member States of a declaration on the basis of Article 24 on the illegal nature of a specific situation cannot be disregarded by them; (2) The scope of Article 25 is not limited to enforcement measures. (3) The wording of these resolutions, for example, the wording of the preamble of resolution 269 (1969) reflects the intention of the Council regarding their obligatory nature."[52]

The preceding considerations lead to the following conclusions: The interpretation of Article 25 of the Charter indicates that Member States

---

[51] ICJ, Reports of Judgements, Advisory Opinions, Orders, 'Legal Consequences for States of the Continued Presence of South Africa in Namibia (South-West Africa), Notwithstanding Security Council Resolution 276 (1970),' Advisory Opinion of 21 June 1970, pp. 53–54.

[52] Sonnenfeld, Renata, 'Resolutions of the UN SC,' 1988, pp. 134–142.

are bound to carry out the provisions of those SC resolutions which are of the nature of a decision and have been adopted in accordance with the Charter. Neither the preparatory work, nor the interpretation based on the provisions of the Charter as a whole proves that the scope of Article 25 should be limited to the provisions of Chapter VII. In the course of its practice hitherto the SC has never conceded that its authority to adopt resolutions binding upon Member States is limited to situations envisaged in Chapter VII. SC resolutions are binding under Article 25, provided that (1) they have been adopted in accordance with the substantial and procedural requirements of the Charter, and no provision expressly envisages the adoption of a recommendation only; (2) the resolution expresses the intention of the SC to bind its addressees.[53]

Following this line, Israel is obligated to follow SC resolutions concerning its nuclear disarmament. Although, the fact remains that according to international law, under the principle of '*pacta sunt servanda*', each State has the sovereign right to become a party to a treaty or not. Israel is nevertheless obligated to join the NPT because, according to the UN Charter, it is obligated to implement and follow the UN SC resolutions.

The SC addressed Israel, as a member of the UN, which has agreed to "accept and carry out" SC resolutions according to Article 25 of the Charter. The claim that the resolution was not adopted under Chapter VII of the Charter or any of the other, previously mentioned Chapters, does not controvert the fact that the resolution was adopted by the Council to maintain peace and security in the Middle East. Hence, SC resolution 487 is mandatory. Also, Israel, under Article 2 of the Charter, should assume its obligations in good faith as a member of the UN that has consented to the Charter. Therefore, taking into consideration, Articles 2 and 25 of the Charter, Israel should have implemented this resolution in good faith during the previous two decades. However, the fact remains that Israel continues to ignore its obligations under the Charter, thereby breaching its obligations under international law.

GA Resolutions

The Israeli nuclear programme and armament has been a subject of annual GA resolutions. Since 1974, the GA addressed the issue of estab-

[53] Ibid, pp. 142–143.

lishing an NWFZ in the Middle East. In resolution 3263 (XXIX), it stated, inter alia, "2. considers that, in order to advance the idea of an NWFZ in the region of the Middle East, it is indispensable that all parties concerned in the area proclaim solemnly and immediately their intention to refrain, on a reciprocal basis, from producing, testing, obtaining, acquiring, or in any other way possessing nuclear weapons ..."[54]

However, subsequent resolutions addressed the question of Israeli nuclear armament directly. They used stronger language condemning Israel for its refusal to renounce any possession of nuclear weapons urging it to place its nuclear facilities under IAEA safeguards. For example, Resolutions A/RES/36/98 of 1981 expressed profound alarm that a report established that Israel has the technical capability to manufacture nuclear weapons and possesses the means of delivery of such weapons. It reaffirmed that the Israeli capability constitutes a serious destabilising factor in an already tense situation in the Middle East and a grave danger to international peace and security. It requested the SC to prohibit all forms of cooperation with Israel in the nuclear field, and to institute effective enforcement action against Israel so as to prevent it from endangering international peace and security by its nuclear-weapon capability. It called upon all States and other parties and institutions to terminate forthwith all nuclear collaboration with Israel. It demanded that Israel should renounce, without delay, any possession of nuclear weapons and place all its nuclear activities under international safeguards. It also requested the SG to give maximum publicity to the report on Israeli nuclear armament and to distribute it to Member States, specialised agencies, the IAEA and non-governmental organisations, so that the international community and public opinion may be fully aware of the danger inherent in Israel's nuclear capability. Furthermore, it requested the SG to follow closely Israeli military nuclear activity and to report thereon as appropriate.[55]

The word "recommendation" has been used in international practice with a variety of nuances and implying a number of effects. Articles 4, 5, 6, and 97 of the UN Charter offer a certain use of the term, where a "recommendation" of the SC is held to be a precondition for a decision of the GA on admission, suspension, or expulsion of States and on the

[54] GA, Resolution 3263 (XXIX), 'Establishment of Nuclear-Weapon-Free Zone in the Region of the Middle East,' 9 December 1974.

[55] GA, Resolution A/RES/36/98, 'Israeli Nuclear Armament,' Distribution General, 9 December 1981.

appointment of the SG. The effect of the SC recommendations was considered at the time of the Corfu Channel case. Discussions in the SC showed an inclination on the part of many delegations to consider that binding obligations could be created by recommendations.[56]

With respect to the GA recommendations under Articles 10 to 14, and even under 58 and 105, neither the Charter nor the San Francisco records indicate what legal effects these recommendations may have. There is no definition of "recommendation." Some delegations conferred an extremely limited authority, while others took a broader view of the authority of the GA. The Charter offers no clarification and the term is used in contemporary practice in many different ways. However, the prevailing view is that "recommendations" of the GA, as distinguished from "decisions" and "declarations," are primarily hortatory. However, even recommendations have legal effects and give rise to certain obligations.[57]

The essential nature of relations between States, in which there is no sharp division between legal and political obligations, has asserted itself within the UN. The fiction that the Court resolves controversies according to law whereas the Assembly and the Council settle political disputes, and therefore, that the resolutions of these organs cannot be sources of law, simply has no validity any longer. Frequently, individual political decisions of the UN organs are based on general legal conceptions. Moreover, these conceptions are sometimes invoked, thus implying an innovative and creative interpretation of the law. Even when these resolutions concern political matters, they are often external manifestations of what an organ of the UN considers the applicable rule of international law.[58]

Furthermore, the GA has given proof, by the terms used in some of its resolutions, that it considers itself competent to participate in its own right in the direct and formal consecration of rules of international law by means of categorical pronouncements about the juridical character of certain practices or of certain principles, without necessarily incorporating them into treaties. Thus, the Assembly has "confirmed" the principles

---

[56] Sloan, Blaine, 'UN GA resolutions in our changing world,' 1991. pp. 26–27. This position was also stated by the then Representative of Australia, Colonel Hodgson, at the 127th meeting on 9 April 1947 in discussing a draft resolution under article 36 of the Charter: "any decision, any recommendation that we may make binds the UK and also binds Albania."

[57] Ibid, pp. 27–28.

[58] Sonnenfeld, Renata, 'Resolutions of the UN SC,' 1988, pp. 4–5.

of the Statute and the judgement of the Nuremberg Tribunal as an expression of international law (Resolution 95 [I]), and it has adopted through the years several "declarations" and other similar pronouncements of a general nature. The adoption of these declarations does not mean that the Assembly creates the juridical norms incorporated in them not that such norms, per se, have a binding nature. The legal value of these pronouncements is not uniform: it is no real creation of norms, there is often legal recognition and confirmation that certain practices or principles are, in the judgement of an organ largely representative of the international community, either customary rules or general principles of international law.[59]

The preceding discussion reveals that some resolutions of international bodies can be manifestations and means of externalisation of international legal norms, that is to say, formal sources of international law. This does not mean, however, that resolutions of international organs ought to constitute an autonomous and distinct source in the same way as treaties, custom, and general principles of law. It is not suggested that Article 38 of the Statute of the Court ought to be modified by adding a new clause relating to resolutions of international bodies. The heterogeneous nature of resolutions does not allow such an act. It is true that treaties, which constitute one autonomous source, also contain dissimilar elements, but contrary to resolutions, all treaties have in common the generation of legal obligations. Neither would the problem be solved by maintaining that only mandatory resolutions, true decisions, constitute a source of law. In the present evolutionary state of this matter, it is difficult to determine the extent of the binding force of certain categories of international resolutions. Moreover, the omission of this source in Article 38 has not prevented the ICJ and other tribunals from recognising, on various occasions, that certain resolutions, whether they have strict legislative character or not, are expressions of law and carry with them obligations in the juridical sense. The reason that is usually given—implicitly invoked, in effect, by the Court in its advisory opinion on reparation for losses incurred in the service of the UN[60]—is that the application of resolutions amounts to the application of international treaties such as the UN Charter, in which the power of the organs to adopt resolutions is

---

[59] Ibid, p. 5.

[60] ICJ, Reports of Judgements, Advisory Opinions, Orders 'Reparation of Injuries suffered in the Service of the UN,' Advisory Opinion of 11 April 1949.

established. Thus, it is said, there is no need, beyond Article 38 of the Statute, for a specific reference to the decisions of international bodies as a source of international law distinct from treaties.[61]

According to the previously stated ICJ Advisory Opinion, the State of Israel, as a member of the UN, should implement the resolutions adopted by UN organs just as it would have to implement international treaties. It should decide unilaterally to disarm and invite the IAEA to verify its compliance with nuclear disarmament. Implementing the UN Charter is in itself a treaty obligation on Israel. Member States are obligated to implement GA resolutions under Article 2 of the Charter. Israel should fulfil its obligations under GA resolutions as well as under SC resolutions in good faith. For political reasons, only mild actions are taken against Israel for ignoring SC and GA resolutions regarding its nuclear armament. However, this does not imply the inadequacy of the international legal rules that obligate Israel to implement such resolutions. Nevertheless, the lack of political will to enforce these resolutions and others are among the reasons behind weakening their implementation.

Legal Effects of GA Recommendations

Duty to Consider. The constitutions of a number of international organisations place a duty on Member States to consider recommendations and to report on action taken. The Charter of the UN does not contain similar provisions, although under Article 64: "the Economic and Social Council ... may make arrangements with the Members of the UN and with the specialised agencies to obtain reports on the steps taken to give effect to its own recommendations and to recommendations on matters falling within the competence made by the GA."[62]

Even in the absence of express provisions in the Charter, there is a duty on the part of Member States to consider a recommendation in good faith and, if requested, to explain their action or inaction. Such obligation is implicit in membership. The classic statement of this duty was made by Judge Sir Hersch Lauterpacht in an often quoted passage from his opinion in the South West Africa Voting Procedure case. Sir Hersch, after noting that a resolution cannot be simply disregarded, stated that a resolution recommending to an Administering State a specific course of action creates some legal obligation which, however rudimentary,

61 Sonnenfeld, Renata, 'Resolutions of the UN SC,' 1988, pp. 5–6.

62 Sloan, Blaine, 'UN GA resolutions in our changing world,' 1991, p. 28.

elastic and imperfect, is nevertheless a legal obligation and constitutes a measure of supervision. The State in question, while not bound to accept a recommendation, is bound to give it due consideration in good faith. If, having regard to its own ultimate responsibility for the good government of the territory, it decides to disregard it; it is bound to explain the reasons of its decision. Judge Klaestad expressed a similar view, in his individual opinion in the same case, by stating that as a member of the UN, the Union of South Africa is in duty bound to consider in good faith a recommendation by the GA under Article 10 of the Charter and to inform the GA with regard to the attitude it has decided to take in respect of the matter referred to in the recommendation.[63]

Since, as seems clear, there is an obligation to consider recommendations in good faith, these obligations are reinforced by Article 2(2) of the Charter which requires all Members to fulfil in good faith their Charter obligations.[64]

Duty to co-operate. It is widely recognised that membership in the Organisation implies at least a duty to consider recommendations. Nevertheless, membership involves more obligations. The ICJ has stressed that membership entails certain mutual obligations of cooperation and good faith incumbent upon the Member and upon the Organisation. While the opinion was expressed in the context of a dispute between Egypt and the WHO over the Regional Official Agreement, the Court made it expressively clear that the obligation to co-operate in good faith is broadly based on the very fact of membership.[65] The Court stated that it considered "those obligations to be the very basis of the legal relations between the Organisation and Egypt under general international law, under the Constitution of the Organisation and under the

[63] Ibid, pp. 28–29.

[64] Ibid, p. 29.

[65] The duty to cooperate has developed through the years. It has gone beyond a factual obligation of membership of an Organisation. In some fields, this duty is considered part of the customary international law, especially in the context of international environmental law. In this regard, Mr. Karayiannopoulos, Fotis A., Barrister and Member of the Board of Appeals of the Community Plant Variety Office, Greece in his Article, "International Environmental Law," stated, "UN GA Resolutions and Declarations have not, in principle, a binding effect. Nevertheless, some strongly support that unanimous resolutions, accompanied by a conform subsequent conduct by the signatory states constitute, at least, a customary rule of law."

agreements in force between Egypt and the Organisation."[66] The Court added, "Moreover, the paramount consideration both for the Organisation and the host State in every case must be their clear obligation to co-operate in good faith to promote the objectives and purposes of the Organisation as expressed in its constitution ..."[67]

The main purpose of establishing an international organisation is to develop international cooperation and, consequently, a basic obligation of membership is to co-operate in achieving the objectives of the organisation. Recommendations are a means for achieving purposes and objectives and, therefore, a duty of membership is to co-operate in carrying out the recommendations.[68] The Charter of the UN supports this thesis. One of the purposes of the Organisation as stated in Article 1 of the Charter is to achieve international cooperation in solving international problems. Moreover, under Article 56, all Members have pledged themselves to take "joint and separate action in cooperation with the Organisation" for the achievement of the purpose set forth in the first Articles of Chapter IX of the Charter. Hence, legal obligations of cooperation embodied in Article 56, as well as those inherent in membership and the good faith element is not only implied but also expressly added by Article 2(2).[69]

The GA certainly has a role in this cooperation since its authority in the field is set out in Articles 13 and 60. Article 60 provides that: "Responsibility for the discharge of the functions of the Organisation set forth in this Chapter [Chapter IX] shall be vested in the GA and, under the authority of the GA, in the Economic and Social Council ..." Article 13 (1) (b), which also deals with economic, social, and

---

[66] Sloan, Blaine, 'UN GA resolutions in our changing world,' 1991, p. 29.

[67] Ibid.

[68] This has been the same view adopted by the GA in its resolution A/RES/59/25, adopted on 17 January 2005 on "Sustainable fisheries, including through the 1995 Agreement for the Implementation of the Provisions of the United Nations Convention on the Law of the Sea of 10 December 1982 relating to the Conservation and Management of Straddling Fish Stocks and Highly Migratory Fish Stocks, and related instruments." It stated, "51. Encourages States fishing for straddling fish stocks and highly migratory fish stocks on the high seas, and relevant coastal States, where a sub-regional or regional fisheries management organisation or arrangement has the competence to establish conservation and management measures for such stocks, to give effect to their duty to cooperate by becoming members of such an organisation or participants in such an arrangement, or by agreeing to apply the conservation and management measures established by such an organisation or arrangement ..."

[69] Sloan, Blaine, 'UN GA resolutions in our changing world,' 1991, pp. 30–31.

human rights matters, authorises the GA to initiate studies and make recommendations to promote international cooperation in these fields. Therefore, it is primarily the GA with which all Members have pledged themselves to co-operate. Emphasising the word "powers" in Article 13 and the pledge in Article 56, it would not be difficult to find more than a right to recommend—even a power of decision—in this text.[70]

Here, however, States have a duty to co-operate with recommendations of the GA. An obligation to co-operate in good faith, however, can not leave a Member free simply to disregard the recommendation.[71]

Duty to Comply. While the duty to consider does not include a duty to comply and the duty to co-operate does not include a duty to accept all recommendations, there are some circumstances in which a duty to comply with some GA recommendations does arise. There is a duty to comply where a State has accepted the recommendation or otherwise agreed to be bound. Such acceptance or agreement may be manifested in various ways (when accepting the recommendation or through a unilateral declaration) and analysis depends on intent. While it is, naturally, the agreement which provides the force or duty to comply, the resolution plays an important role in achieving or reflecting the agreement.[72]

Where a State has accepted a recommendation, its duty to comply is clear. There may, however, be circumstances in which, in the absence of an actual acceptance, a State may still, as a result of its affirmative vote or even its acquiescence, or other actions before or after the vote, be bound to comply with a recommendation. If the conduct of a State gives rise to reasonable expectations on the part of other States and if the other States have acted upon these expectations, a State may be estopped or precluded from denying an obligation to comply. A further situation in which a duty to comply may be inferred involves the question of abuse of right. Judges Lauterpacht and Klaestad, in recognising a duty to consider a recommendation, observed that such duty did not involve a binding legal obligation to comply. However, Judge Lauterpacht carried the analysis further and suggested that repeated failure to comply might result in an abuse of right.[73]

---

[70] Ibid, p. 31.
[71] Ibid.
[72] Ibid, pp. 32–33.
[73] Sloan, Blaine, 'UN GA resolutions in our changing world,' 1991, pp. 34–35.

When the obligation to co-operate in good faith is ignored to the point of becoming an abuse of right, a breach of the duty to act in good faith emerges. At some stage in the process, compliance will become necessary to avoid an abuse of right.[74]

Basis for Implementation or Enforcement. The GA may take a number of implementing measures short of enforcement action. It may adopt further resolutions calling on a State to comply with its recommendation or condemning it for failure to comply. It may call for information and reports on implementation from the SG or from Member States.[75]

Moral and Political Effects. Apart from the question of legal effects, GA recommendations have persuasive effects. It is not easy, however, to measure the hortatory, political, or moral force of a recommendation or to define the nature of such force. Nor is there any agreement on even the meaning of these terms. A resolution by the GA, an international political body, invariably has some "political effect". However, this does not entail a political obligation since political science is a descriptive and not a regulatory or normative science.[76]

Professor I.I. Lukashuk,[77] of the Institute of State and Law, USSR Academy of Science, wrote: [I]n strict legal terms all resolutions of the GA have the same status and are not legally binding. However, such status does not mean these acts may not possess a binding force of a different kind. The lack of any binding force would make the resolutions senseless, and the UN would have lost an important instrument for influencing international relations, so since the resolutions are called upon to influence both the consciousness of a State and inter-State relations directly ... Under its Charter, the UN must be a centre for harmonising the actions of nations in the attainment of common ends. Resolutions are an important mean to this end. The ICJ has confirmed that the very fact of UN membership presupposes certain mutual obligations of cooperation and good faith,[78]

---

[74] Ibid, p. 36, Professor Gregory I. Tunkin has concluded that since recommendatory resolutions are one means for developing cooperation and achieving the objectives of the international organisation, a persistent disregard of such recommendations by a Member State would constitute a violation of the organisation's Charter.

[75] Ibid, pp. 37–38.

[76] Ibid, p. 43.

[77] Ibid, pp. 43–44.

[78] ICJ, Reports of Judgements, Advisory Opinions, Orders, 'Interpretation of the Agreement of 25 March 1951 between the WHO and Egypt,' Advisory Opinion of 20 December, 1980.

imposed on the State and the Organisation. He concluded that GA recommendations have moral and political binding force. However, the concept of binding political force, like the concept of soft law, is elusive. One suggested distinction between a legal and a political obligation is that legal obligations may be enforced in judicial proceedings while political obligations may only be asserted in a political body, or perhaps in negotiations. Lukashuk, who denies any legal force to a recommendation, seems to explain the basis for political force by reference to obligations of cooperation and good faith. The ICJ in the Egypt-WHO advisory opinion supported these obligations. He also refers to "abuse of right" which is a legal concept.[79]

The use of legal concepts in referring to political obligations entails an inherent contradiction between what is political and what is legal. The political effect of any resolution undertaken by UN bodies was never challenged, as it represents the political will of the majority of nations. The problematic issue lies in the legal effect of GA recommendations. The Charter does not imply that there should only be moral effect to the GA resolutions, although it did not clarify that such recommendations are mandatory like SC resolutions. However, such recommendations have legal value and member States should undertake to fulfil them in good faith according to the Charter.

"Soft Law". The concept of "soft law," a term, whose origin is attributed to Lord McNair,[80] is related chiefly to three main categories: (1) so called "non-binding" agreements, of which the Helsinki Accords[81] have been the centre of controversy; (2) "voluntary" codes of conduct for transnational corporations, and (3) resolutions of international organisations, of which GA resolutions are the leading example.[82]

The effects of "soft law" are difficult to measure. The term is of doubtful usefulness in considering the effects of GA resolutions. In any event, it cannot be applied generically in any of its varied meanings to all resolutions. Some resolutions will include "hard law." The provisions of other resolutions, like the provision of some treaties, may be "soft law" in the sense of being vague and imprecise, while other provisions may be

---

79 Sloan, Blaine, 'UN GA resolutions in our changing world,' 1991, pp. 43–44.

80 Ibid, p. 50.

81 The Helsinki Accords is the Final Act of the Conference on Security and Cooperation in Europe held at Helsinki in 1975.

82 Sloan, Blaine, 'UN GA resolutions in our changing world,' 1991, p. 50.

clearly and precisely drafted. For those who accept degrees of obligation or legal force, there will be resolutions which are fully binding and others where the force, though real, is something less. In some resolutions the obligations are limited to those involving good faith and cooperation. Perhaps a real significance for the present subject rests in the idea of soft law as a law in transition. Resolutions, even recommendatory resolutions, are in the process of becoming something more. They may, in the words of Professor Dupuy, affirm a legitimacy which anticipates the legality of tomorrow.[83]

Considering the previous analysis, GA resolutions concerning Israeli nuclear armament are of certain legal value. The GA repeatedly requested Israel to cease from developing nuclear weapons and to place its nuclear facilities under full-scope safeguards. It has requested all Member States to co-operate towards achieving this goal. Determining the legal value of these resolutions is not an easy task. However, Israel has a general duty to comply and to act in good faith to implement GA resolutions as a member of the UN. The duty of membership to the UN requires States to co-operate in carrying out recommendations that serve the object and purpose of the Charter. The continuation of the threat that Israel may use its nuclear weapons in any conflict in the region constitutes a violation of its obligations under Chapter VII of the Charter, which implies that all Member States should refrain from the threat or use of force in their relations. In addition, the UN Charter is sufficient to provide guidance relating to the application of decisions and recommendations of the UN bodies. Under Articles 1 and 2, all Member States are requested to co-operate to fulfil their obligations in order to fulfil the object and purpose of the Organisation.

Moreover, Israel completely disregards these resolutions; it does not even consider them, which constitutes a breach to its Charter obligations to co-operate with the organisation to establish peace and security, which are the main goals of the UN. This situation entails an abuse of right by Israel, i.e., Israel is abusing its UN membership rights by not complying with or considering the implementation of GA resolutions regarding its nuclear armaments and by not co-operating to establish an NWFZ in the Middle East. In addition, the GA recommendations are considered a soft law, or represent moral and political statements, they still have a legal value under the UN Charter. Therefore, a State should not abuse its right

[83] Ibid, p. 51.

as a member of the UN and should work towards complying with its obligations under the Charter. Therefore, a state like Israel should not persist in its non-compliance with GA resolutions as long as they were undertaken within the GA competence and following the appropriate procedures indicated in the UN Charter and they serve the main goals of the Organisation.

3. *Conclusions*

According to general rules of international law, a State is only bound by a treaty it ratifies. However, the law does not specify what should happen in a situation where the UN organs call upon a certain State to join a certain treaty. Therefore, in the absence of a rule dealing with this matter, and without jeopardising the State's sovereign right to join a treaty or not, the addressee is still under obligation to fulfil its legal obligations to implement the resolutions of the UN organs.

By maintaining its position outside the NPT, Israel is not bound by its rules. However, according to the UN Charter, the State of Israel is under legal obligations to implement the resolutions adopted by the SC and to act in good faith to carry out the decisions of the GA as well, which have called upon it to halt its nuclear military activities and place its nuclear facilities under IAEA safeguards. However, the fact remains to this day that the State of Israel has not implemented either SC or GA resolutions.

If Israel keeps its nuclear arsenals for more security, the 1973 war laid to rest a number of myths about the Israeli nuclear deterrent. For one thing, it brought into question whether the A-bomb can be regarded as an effective deterrent against massive Arab conventional aggression. It is obvious that the possibility of nuclear retaliation did not deter Egypt and Syria from invading Israel in 1973—even if it may have influenced the limiting of their war aims.[84]

The problem of the rules is engraved in the NPT "haves" and "have-nots" system. Israel, like India and Pakistan, refused the current rules because of its discriminatory nature. The NWS NPT members did not fulfil their obligations under Article VI of the Treaty. Thirty five years since the NPT entered into force in 1970, NWS continue to develop their nuclear weapons instead of halting activities to produce fissionable materials or

[84] Hersh, Seymour M., 'The Samson Option: Israel's Nuclear Arsenal and American Foreign Policy,' 1991, p. 116.

the manufacture and development of new nuclear explosive devices and this is the reason behind the current situation. The lack of a system to monitor both the implementation of NPT rules by NWS and the transfer of nuclear technology among them or between them and non-NPT members represent an inadequacy of the rules. France co-operated with Israel to build Dimona when France was not an NPT member; but sold Osirak to Iraq when they were both NPT members.

This loophole in the rules leads to such problems and eventually the international community should consider adopting rules to deal with this gap. States should be held accountable for their improper implementation of the rules.

The obligation on Israel is derived from its obligations as a member of the UN. If the rule of "*pacta sunt servanda*" is applied here, Israel, by joining the UN, is bound to enforce UN resolutions. This obligation is clearly stated in the UN Charter. Under the general rules of customary international law and the Charter, Israel should refrain from the threat or use of force, through conventional or nuclear weapons, against other members of the international community. Nevertheless, if, strategically speaking, the balance of power in the Middle East region requires the acquisition of conventional weapons, there has so far been no need in the area for the existence of WMD, especially nuclear weapons. The existence of nuclear weapons, as such, constitutes a threat against other States in the region. Israel can be considered in breach of its international obligations under customary international law and the UN Charter. The dilemma, in this case, as in most other international cases, is the inherent contradiction between the lack of political will and the enforcement of legal obligations. Hence, if the political will exists, enforcement of legal obligations take place, as in the case of Iraq. However; the lack of political will leads to the disregard of a situation where a State is in abuse of its membership rights, as in the case of Israel. The problem is not in the rules, but in their process of enforcement through political bodies that are governed by the political interests of their members.

CHAPTER SEVEN

ESTABLISHING AN NWFZ IN THE MIDDLE EAST

I can smell the wind of change blowing in the air. (NN)

1. *Introduction*

The security tensions in the world, especially in the Middle East, give rise to the dangers caused by the introduction of nuclear weapons in the region. After Israeli nuclear capabilities came to the attention of States in the region, they adopted different stances. On the one hand, Iraq worked hard for acquiring military nuclear capabilities, so as to introduce counter deterrence against Israeli nuclear weapons. On the other, Iran, in 1974, introduced a proposal to call for the establishment of an NWFZ in the Middle East. It was supported by Egypt and other States in the region.

In the region, this proposal has been received over the years with both caution and acceptance. Even Israel, the only NWS in the region, has accepted the proposal. However, there are differences among States in the region concerning the steps required to achieve the end goal of the proposal, i.e., establishing the NWFZ.

Israel associates negotiations concerning nuclear disarmament with developments in the peace process. It maintains that the establishment of an NWFZ has to be negotiated among States in the region. However, other States in the region view that such negotiations are not imperative for the establishment of an NWFZ and that it is not necessary to associate the nuclear issue with the peace process. All other States in the region are members of the NPT and their nuclear facilities are placed under IAEA safeguards. Hence, Israel is thereby invited to join the NPT as an NNWS and to place its nuclear facilities under IAEA safeguards. This will ensure that its programme remains peaceful, similar to other programmes in the region, while at the same time will work towards achieving the goal of establishing an NWFZ in the Middle East.

In this chapter, the history of efforts to establish an NWFZ in the Middle East will be discussed.

2. *History of Attempts to Establish an NWFZ in the Middle East*

After the idea of establishing an NWFZ in the Middle East captured the attention of international organisations and the different international forums dealing with nuclear disarmament in 1974, it became a long-standing issue in the annual GA meetings and the subject of several studies and resolutions. It was again one of the major issues of the 1995 Extension and Review Conference of the NPT, where a separate resolution was issued specifying the importance of establishing this zone. This position was maintained in the work of the 2000 Review Conference of the NPT and again during the work of the Preparatory Committee for the 2005 Review Conference of the Treaty.

The 1974 GA 3263 (XXIX) Resolution commended the idea of establishing an NWFZ in Middle East. It declared that States in the area "should refrain, on a reciprocal basis, from producing, testing, obtaining, acquiring or in any other way possessing nuclear weapons and called upon them to accede to the NPT."[1]

The said Resolution which was adopted in the 1995 Review and Extension Conference called upon States in the Middle East to accede to the NPT as soon as possible and to place their nuclear facilities under full-scope IAEA safeguards.

The 2000 Review Conference reaffirmed the importance of the Resolution. It recognised that the resolution will remain valid until its goals and objectives are achieved. The Resolution, which was co-sponsored by the depository States (Russia, the UK and the US), is an essential element of the outcome of the 1995 Conference and of the basis on which the NPT was indefinitely extended without a vote in 1995. The resolution also noted that all States in the Middle East, with the exception of Israel, are party to the NPT and reaffirmed the importance of Israel's accession to the NPT and the placement of all its nuclear facilities under comprehensive IAEA safeguards, in realising the goal of universal adherence to the treaty in the Middle East. It requested States of the region to transmit declarations of support to the SG of the UN and to take practical steps towards the objective of establishing an effectively verifiable NWFZ in the Middle East as well as other WMD. It also appealed to all NPT States to cooperate and to exert their utmost efforts to ensure the early establishment by regional parties of a Middle East zone free from

[1] GA, RES 3263 (XXIX), 1974.

WMD and their delivery systems. It noted the commitment by the five NWS of the 1995 Resolution on the Middle East.[2]

3. *Concept of Establishing an* NWFZ

On 11 December 1975, Resolution 3472 (XXX) of the GA defined an NWFZ as "any zone, recognised as such by the GA of the UN, which any group of States, in the free exercise of their sovereignty, has established by virtue of a treaty or convention." Certain countries, such as Pakistan, have rejected the argument that the prior agreement of all States of the region concerned is essential for the creation of an NWFZ, a condition which has only served to delay the establishment of an NWFZ in Africa, the Middle East and South Asia. Such projects should be encouraged in spite of the reservations of one or two States in the regions.[3]

This argument could be valid in certain cases if the one or two States objecting to the establishment of an NWFZ are NNWS. However, the establishment of such a zone with the objection of a NWS in the concerned region would be basically insignificant. If the threats caused by one or two States in a region were to remain, the advantage of establishing an NWFZ in any specified region would be uncertain.

The GA decided that the instrument establishing an NWFZ should provide for:

- The statute of total absence of nuclear weapons to which the zone shall be subject, and,
- An international system of verification and control to guarantee compliance with the obligations deriving from that statute.

The resolution contains no precise requirements as to the territorial scope of an NWFZ. In fact, the scope of a zone depends on political, strategic and geographical considerations. In general, an NWFZ is made up of the land, air and maritime territory of several neighbouring States which conclude an agreement to this end. However, there is nothing to prevent a single State from turning part or the whole of its territory into an NWFZ. Thus, several situations are possible:[4]

---

[2] NPT Review Conference, NPT/CONF.2000/28, 'Final Document of the 2000 Review Conference of the Parties to the NPT,' 24 May 2000, pp. 20–21.

[3] Biad, Abdelwahab, 'Nuclear-Weapon-Free Zones: Challenges and Prospects for Non-Proliferation,' Nuclear Law Bulletin, Issue 58, 1996, p. 16.

[4] Ibid, p. 17.

- A State belongs to an NWFZ but its overseas territories do not;
- A NWS has a military base on the territory of another State located within an NWFZ.
- A NWS has sovereignty over territories located within an NWFZ.

3.1. *Obligations of the States Party to an NWFZ*[5]

The States Party to an NWFZ undertake to use nuclear energy exclusively for peaceful purposes and, in order to ensure compliance with this undertaking, set up a system for controlling activities.

Non-Acquisition of Nuclear Weapons

States in an NWFZ are prohibited from:

- developing and manufacturing nuclear weapons;
- possessing such weapons in any way, including storage;
- acquiring nuclear weapons in any manner;
- conducting nuclear tests in the zone;
- deploying nuclear weapons on the territory of States party to the adjoining maritime areas.

Control of the Uses of Nuclear Energy

The purpose of control is to ensure that nuclear energy is used for exclusively peaceful purposes. The NWFZ treaties have a system of control, usually based on the following elements:

- Application of IAEA safeguards.
- Reports and exchange of information.
- Complaints procedure.

Obligations of NWS

The obligations of the NWS as regards NWFZ are that NWS are required not to introduce nuclear weapons into the zones and not to use or threaten to use such weapons against the Parties through "positive" and "negative guarantees".

---

[5] Ibid, pp. 20–24.

3.2. *Positions of Involved Parties Towards Establishing an* NWFZ *in the Middle East*

Before detailing the position of the involved parties towards establishing an NWFZ in the Middle East, it is worth mentioning that this initiative is supported by the international community as expressed in GA and SC resolutions on the matter. It should be noted that these resolutions were adopted by consensus.[6]

As for the involved parties' positions, both Israel and the Arab countries support the initiative to establish an NWFZ in the Middle East. Nevertheless, there is a fundamental difference in approach.[7]

Arab Countries and Iran's Position

The Arab countries and Iran initiated and sponsored all the resolutions adopted by the GA and the NPT Review Conferences to establish an NWFZ in the Middle East.

Realising the political situation in the region, these States focus their efforts on requesting that all States in the region join the NPT and remain in compliance with their obligations under IAEA/NPT safeguards. Considering that all the Arab States in the region and Iran are members to the NPT and their facilities are under IAEA safeguards, Israel is the only State requested to join the Treaty and place its facilities under IAEA safeguards.

States in the region, other than Israel, consider that if the latter joins the NPT, this will in itself entail the establishment of an NWFZ in the area. They consider it a step forward towards non-proliferation of WMD in the region.

Hence, the Arab States and Iran reckon that due to the current political situation, it is more practical not to resort to negotiations to establish an NWFZ in the region. The Arab states and Iran endorse the steps that are

---

[6] Marom, Ran, 'Israel's Position on Non-Proliferation,' June 1996, also in the CoB of NAM letter, it was stipulated "The Coordinating Bureau reiterated its support for the establishment in the Middle East of a zone free of all weapons of mass destruction. To that end, the Bureau reaffirmed the need for the speedy establishment of a nuclear-weapons-free zone in the Middle East in accordance with SCR 487 (1981), and paragraph 14 of SCR 687 (1991), and the relevant GA resolutions adopted by consensus, and stressed that necessary steps should be taken in different international forums for the establishment of the zone."

[7] Barnaby, Frank, 'Capping Israel's Nuclear Volcano,' in Efraim Karsh (ed.), 'Between War and Peace: Dilemmas of Israeli Security,' 1996, pp. 109–110. This was the same view Stated by Marom, Ran, 'Israel's Position on Non-Proliferation,' June 1996, pp. 57–59.

stipulated in the GA resolutions to establish this zone. These steps request the only nuclear State in the region—Israel—to join unilaterally the NPT, halt the production and development of its nuclear arsenal and place its nuclear facilities under full-scope safeguards. However, consecutive Israeli governments have objected to these steps.

Israeli Position

Israel abstained from voting on draft resolutions adopted annually from 1974 to 1979 by the GA, due to reservations regarding their wording and for reasons related to the modalities proposed by their sponsors. Israel pointed out that these modalities were neither in accordance with international general practice nor in conformity with the precedent of the Treaty of Tlatelolco. They were considered by Israel "as not practical and self-defeating, since without full and frank consultations between all States concerned, conducted in an atmosphere of mutual trust and goodwill, the proposal could not be realised." Moreover, "in the absence of a negotiated agreement, it is evident that a mere series of unilateral proclamations ... would be nothing but a futile exercise in rhetoric, without any practical or legally binding meaning." The issues involved in the operative paragraphs of those resolutions were exceedingly complex and such a matter as the establishment of an NWFZ could not "be settled by correspondence through the SG" or by "unilateral depositions of declaratory statements of intent," or by a mere series of unilateral acts.[8]

This idea is somehow awkward; correspondence through the SG is not actually necessary. If Israel joined the NPT as an NNWS, the zone would thence be free from nuclear weapons.

In 1975, Foreign Minister Yigal Allon stated that "Israel supports the proposal for an NWFZ in the Middle East and will be ready to enter into negotiations with all States concerned in order to attain that objective." In 1976, Israel called upon its Arab neighbours "to sit together without any prior conditions and discuss peace; if this is not acceptable to them we are prepared ... to discuss arms control and the proper conditions for the creation of an NWFZ in the Middle East, independent of an immediate settlement." Furthermore, in its note of 14 September 1976 to the UN SG, "Israel reaffirms its readiness to participate in a conference

[8] Marom, Ran, 'Israel's Position on Non-Proliferation,' June 1996, pp. 57–58.

of all States of the region and to negotiate forthwith the establishment of an NWFZ in the Middle East." In 1977, Foreign Minister Moshe Dayan repeated this invitation and since 1979, after signing the peace treaty with Egypt, Israel has proposed that "direct negotiations with the participation of all the States in the region could start without preconditions at any time and in any place." Israel, for its part, stated that it "stands ready to begin such negotiations without delay or preconditions." Foreign Minister Yitzhak Shamir reiterated this approach in a Knesset debate on 30 July 1980 and twice at the UN GA in its 38th and 40th sessions of 1983 and 1985.[9]

At the 35th session of the GA of 1980, Israel introduced a draft resolution, the operative paragraph of which "1. Calls upon all States of the Middle East and NNWS adjacent to the region, that are not signatories to any treaty providing for an NWFZ, to convene at the earliest possible date a conference with a view to negotiating a multilateral treaty establishing an NWFZ in the Middle East; 2) Urges all States of the region to state by 1 May 1981 their willingness to participate in the conference." Israel maintained that its "offer to the Arab States of the region and to States adjacent to the region is an unlinked deal." It appealed to them since "regardless of their political differences and without prejudice to any political and legal claim, these States must be prepared to take concrete steps toward the achievement of this vital objective." Most Arab States rejected this proposal and as a result Israel withdrew its draft resolution.[10]

In June 1981, Israel further elaborated its proposals on the matter. It "formally and urgently turns to all States of the Middle East, and to States adjacent to the region, and requests that each indicated in the course of 1981 their consent to the holding of a preparatory conference to discuss the modalities of such a conference of States of the Middle East, with a view to negotiating a multilateral treaty establishing an NWFZ in the Middle East." Furthermore, taking note of an Egyptian proposal aired in April 1981 suggesting the preparation of a study on the modalities for establishing such a zone—Israel, on its part, proposed that "the study be undertaken by qualified experts from the Middle East States, including Israel." In its opinion, there was "no need to wait until all Governments in the Middle East see their way to endorsing such a study: undertaking it in a manner proposed by Israel would, by itself, constitute a valuable

[9] Ibid, pp. 61–62.
[10] Ibid, p. 62.

step in the direction of building much-needed confidence between the States of the region prior to the holding of the preparatory conference proposed above."[11]

At the 36th session of the UN GA held in 1983, Israel again called upon "all the States of the region, for the sake of their common future, to take concrete steps towards the establishment of an NWFZ in the Middle East." As far as Israel was concerned, on 18 June 1982, Prime Minister Begin stated in his address to the 12th Special Session of the UN GA that "Israel is prepared to negotiate and sign such a treaty [patterned on the Tlatelolco Treaty] with all its neighbours in the Middle East."[12]

This approach led Israel to express reservations about the modalities envisaged by the Egyptian draft resolutions introduced at the 35th session of the GA (1980), and it has continued to do so annually since the 37th session (1982). During each of these sessions, Israel maintained that the Egyptian draft resolution, though designed to attain the same aim, neglects mention of the negotiating process, without which the Treaty for the establishment of an NWFZ in the Middle East is unlikely to come about. Nevertheless, because it wished to signify its support for the establishment of such a zone, it joined the consensus on this issue during the above-mentioned sessions.[13]

From the Israeli perspective, the task of averting nuclear proliferation in the Middle East makes it imperative upon all States in the region to overcome their political differences and to initiate negotiations aimed at working out relevant collective measures. Such negotiations should be carried out regardless of political solutions of disputes and conflicts. The promotion of any attempt to effectively implement such a scheme would depend upon an agreement by States in the area to several principles:

1. The carrying out of direct negotiations without preconditions and in good faith.
2. The formulation of a treaty defining the obligations and rights of the contracting parties, establishing the machinery for ensuring compliance of States with it and other issues agreed to by the negotiating States.
3. A flexible approach to the definition of the zone.

---

[11] Ibid, pp. 62–63.
[12] Ibid, pp. 63–64.
[13] Ibid, p. 64.

4. The acceptance by all States of the provisions of the UN Charter and other accepted norms of international law as standards for conduct of relations.

3.3. *Situation in the Middle East*

The Persian Gulf and the Middle East, nominally separate geographic identities, are sometimes regarded as one broad political-military region. The dynamics of the Arab-Israeli conflict, and Israel's relative military superiority, invariably affect the thinking of all the Arab/Persian communities in the region. Even without Israel, there are inter-Arab and regional rivalries that provide impetus for proliferation. The Iran-Iraq war was a powerful formative experience for both these countries and their concept of national security and deterrence. Likewise, Israel's long history of conflict with its neighbours, its innate sense of vulnerability, and the hostility it faces from the Gulf States continue to nourish its own elusive search for security.[14]

Some authors argue that even though Israel already enjoys a clear military superiority over its Arab adversaries, singly or combined, acquisition of nuclear capability would greatly enhance Israel's sense of security. In this circumstance, certain Israelis might be inclined to adopt a moderate and conciliatory posture, they believe it much more likely, however, that Israel's policy toward its neighbours would become more rather tough than less. Israel would seek to exploit the psychological advantages of its nuclear capability to intimidate the Arabs and to prevent them from generating trouble on the frontiers. Others have hypothesised that Arab acquisition of the bomb would make the Middle East more stable by creating a "balance of terror". Shai Feldman, for instance, develops this argument. He states that "By providing effective deterrence, nuclear weapons have the capacity to reduce the likelihood of war, both conventional and nuclear."[15] However, in Catudal's view, "the proliferation of nuclear weapons in the Middle East would be a dangerous development. They are destabilising. Crisis situations might tempt both sides to strike first in the hope of landing a first strike. They also require parties to keep their forces on a "hair trigger" alert."[16]

---

[14] Donovan, Michael, 'Iran, Israel and Nuclear Weapons in the Middle East,' 14 February 2002.

[15] Feldman, Shai, 'Israeli Nuclear Deterrence, A Strategy for the 1980s,' 1982.

[16] Catudal, Jr., Honore M., 'Israel's Nuclear Weaponry—A New Arms Race in the Middle East,' 1991, p. 119.

In fact, Israeli nuclear deterrence did not prevent it from being attacked in the 1973 war. It did not stop Iraq from bombing Israel during the 90s and it did not prevent the troubles on its frontier with Lebanon. At the same time, the creation of a "balance of terror" may not be a factor of stability in the region. It is still boiling in turmoil. In addition, the balance that other States in the area may seek, while they are not developing nuclear weapons, is to build and accumulate a stockpile of chemical and biological weapons thereby creating increasing dilemmas for the region.

Israel's destruction of the Osirak reactor outside Baghdad in 1981 demonstrates that the protection of its regional monopoly in fission weapons may be a definite component of Israel's nuclear policy. While Tel Aviv is determined to prevent the Arabs from "going nuclear", this is probably an impossible task. Every major Arab State now conducts some activities in the nuclear field. Some have devoted their efforts to a slow build up of a nuclear infrastructure; others have attempted "a quick fix" acquisition of nuclear weapons.[17]

It is not accurate to accept this statement. The States in the Middle East, not taking into consideration failed attempts by Iraq and Libya, are building up their nuclear infrastructure in a transparent manner. These facilities are for peaceful uses and placed under IAEA safeguards. Most of them are research reactors and few are devoted to produce energy (including Iran). The States in the region do not even depend on nuclear power to generate electricity. Considering all this, it is not possible to state for a fact that there is a build-up, whether slow or not, of a nuclear infrastructure in the Arab States.

Despite the statement of Israel's Prime Minister listing Israel among NWS, Israel's official position adopts a policy of nuclear ambiguity, which states that it will "not be the first to introduce nuclear weapons into the region." However, Israel developed nuclear weapons to be a deterrent and a last resort weapon to ensure that the State of Israel will never be demolished. The explanations of Israel's nuclear-weapon force do not justify its size and quality. A last-ditch deterrent, a deterrent against attacks on Israeli cities by chemical or biological weapons and an insurance against loss of technical superiority would require no more than 30 or so ordinary fission nuclear weapons, with explosive powers of between 20 and 40 Kt, enough to destroy the major cities of its Arab adversaries. A larger

[17] Ibid, p. 118.

nuclear weapon force and weapons with larger explosive yields or of more sophisticated design are not necessary for these purposes.[18]

It is hard to conceive of any legitimate tactical or other use for Israeli nuclear weapons. The harsh reality is that the country's area is that the fallout from any detonation of a nuclear weapon in the area would be a major hazard for Israeli civilians as well as soldiers. In March 1961, Dr. Nahum Goldmann, who served as President of the World Zionist Organisation and the World Jewish Congress, participated in a symposium celebrating the 13th Anniversary of Ma'ariv, a major Israeli daily newspaper, stated that "nuclear weapons would not improve Israel's security situation." Moreover, he argued that no country had "the moral right to produce an atomic bomb." A CIA internal memo, prepared by Sherman Kent, Chairman of the Board of National Estimates at the Central Intelligence Agency in March 1963, defined the "acquisition" by Israel to be either (a) detonation of nuclear device with or without the possession of actual nuclear weapons, or (b) an announcement by Israel that it possessed nuclear weapons, even without testing. In general, the thrust of Kent's memo was negative. He said that an Israeli bomb would cause "substantial damage to US and Western position in the Arab World."[19]

Although some reports indicate that Israel instituted a nuclear alert during the 1973 War and again in 1991 during the Gulf conflict, information about the Israeli weapons program is somewhat conjectural. The Israeli Government does not admit to possessing nuclear weapons and is not a member of the NPT. Dimona remains a closed site not subject to international inspections or safeguards. There exists no official mention of how nuclear weapons fit into Israeli strategic thinking, and their role in the Israeli Defence Force's doctrine is therefore a matter of guesswork.[20]

Israel needs nuclear weapons to pre-empt enemy State nuclear attacks. Should Israel ever need to actually use its nuclear weapons for such a purpose would signify the failure of these forces as a deterrent. Such failure is increasingly plausible because of the complex nature of nuclear deterrence in general and because of the unique context of the Middle

[18] Barnaby, Frank, 'Capping Israel's Nuclear Volcano,' in Efraim Karsh (ed.), 'Between War and Peace: Dilemmas of Israeli Security,' 1996, pp. 106–107.

[19] Catudal, Jr., Honore M., 'Israel's Nuclear Weaponry—A New Arms Race in the Middle East,' 1991, pp. 16–35.

[20] Donovan, Michael, 'Iran, Israel and Nuclear Weapons in the Middle East,' 14 February 2002, p. 3.

East in particular.[21] The hypothesis hence would be; if Israel will never require to actually use nuclear weapons against its enemies, then keeping them as a deterrent will not be a guarantee for its security either.

The Israeli nuclear weapons were intended to slow down Arab efforts to develop a countervailing nuclear capability, but it was only partially successful. Egypt, Israel's formidable Arab adversary in the past, quietly acquiesced. Syria developed a significant chemical capability as a cheap substitute for a nuclear deterrent. Iraq, defying the "Israeli diktat," pushed relentlessly ahead with a massive under-cover effort to develop a variegated and exceptionally well-protected nuclear programme. Had it not been for Saddam's miscalculation on 2 August 1990, this would probably have earned Iraq a nuclear capability by now.[22]

Whether or not the Iraqi programme would have resulted in enhanced mutual deterrence is highly questionable, but for all intents and purposes, this is a speculation now that Iraq has been disarmed of all its WMD.

The obvious conclusion of the preceding discussion is that Israel's only *de facto* long term hope for a stable equilibrium in the Arab-Israeli arms race lies in a regional system of arms control in which the ability of the Arabs to acquire non-conventional weapons, as well as a mass conventional firepower, is curbed by third party (especially great powers) intervention. Such a system cannot ante cede a multilateral settlement of the Arab-Israeli conflict. But it ensures that, if and when such a settlement is worked out, the Israelis will ultimately agree to some form of inspection and control over their own nuclear programme.[23] So far, decades after Israel became an NWS, it has not yet accepted either inspection or any kind of control to limit its nuclear programme.

Efforts to force Israel to surrender the nuclear arsenal have gathered momentum over the past few years.[24]

To those who have targeted Israel in this area of acute sensitivity, the nuclear programme of the Jewish State represents the source of all evil, the original sin. In their view, it was Israel's decision to go nuclear that

---

[21] Beres, Louis Rene, 'Israel's Bomb in the Basement: A Revisiting of "Deliberate Ambiguity" vs. "Disclosure,' in Efraim Karsh (ed.), 'Between War and Peace: Dilemmas of Israeli Security,' 1996, p. 121.

[22] Yaniv, Avner, 'Non Conventional Weapons and the Future of Arab-Israeli Deterrence,' in Efraim Karsh (ed.), 'Between War and Peace: Dilemmas of Israeli Security,' 1996, pp. 138–139.

[23] Ibid, pp. 148–149.

[24] Karsh, Efraim and Martin Navias, 'Israeli Nuclear Weapons and Middle East Peace,' in Efraim Karsh (ed.), 'Between War and Peace: Dilemmas of Israeli Security,' 1996, p. 75.

enticed the Arabs to the verge of an Armageddon. Some would even go so far as to regard Israel's nuclear programme as the driving force for non-conventional proliferation in the entire Third World, and a major obstacle to efforts to expand global membership of the NPT and to ensure adherence to its provisions. However, the opposite view raises qualms based on the difficulty of finding even one non-conventional weapons project in the Middle East as a reaction to Israel's nuclear weapons.[25]

The first view is more accurate. The historical link between Arab and Israeli nuclear programmes has not been exaggerated. Nuclear military efforts in the region started in the 70s after the realisation of the Arab States that Israel turned into a NWS. Nevertheless, the fact remains that in today's world, the Arab States are NNWS and that Israel's insistence in developing its nuclear arsenals is enough of an incentive to these States to develop other WMD or to aspire to build up nuclear military capabilities which would stand as a deterrent against the Israeli nuclear threat. Iraq and Libya are good examples.

With little stake in preserving an inequitable system of this kind, a number of third world States deliberately set about to circumvent the regime. India and Israel began their nuclear programmes in the 1960's, for example, and were assisted by industrial States that put commercial advantage ahead of their stated non-proliferation objectives. In fact, circumvention of NPT controls was facilitated by transfers from the advanced nuclear power States, including parties to the NPT. The US, Canada, and the UK helped India in various ways; France and the US assisted Israel; and Germany, France and the then USSR assisted Iraq. Only recently have Third World States who are not adherents to the agreement also become nuclear suppliers; these States include China and possibly Israel.[26]

The global proliferation of nuclear weapons is the responsibility of all the proliferators. Each States may argue that it is not the only proliferator, so it might be appropriate to disarm the others first. It is true that each of the previously mentioned proliferators had their own regional tensions that prompted them to develop nuclear weapons, nevertheless, this does not mean that Israel's nuclear weapons have nothing to do with it. Israel supported the South African apartheid regime to develop nuclear capabilities. They would not have achieved it as fast as they did without the

[25] Ibid.

[26] Nolan, Janne, 'Technology and Non-Proliferation in a Changing World Order,' Transnational Law and Contemporary Problems, Volume 2, Number 1, 1992, p. 407.

support and science of Israeli nuclear specialists. Moreover, Israel today is cooperating with India in different nuclear fields including military ones. So, it is inappropriate to say that Israeli nuclear proliferation does not hamper efforts to bring universality to the NPT and to guarantee that more nuclear proliferators will not.

The States arrayed against Israel hold that it is their right to develop nuclear weapons as a deterrent to the Israeli arsenal. They believe that Washington maintains a double standard by ignoring Israel's acquisition of WMD while opposing the transfer of even peaceful nuclear technologies to others. Both Iraq and Iran have sought a nuclear capability as a strategic equaliser. In the case of Iraq, however, nuclear weapons served an ambition greater than that of a relative deterrent. Saddam's search for regional hegemony meant that he was both to overcome Iran's strategic superiority and to stake a claim for leadership of the Arab world. In Saddam's eyes, WMD in general and nuclear weapons specifically, served both purposes. They countered Iran's strategic depth and demographic superiority. They were also seen to threaten and confront Israel by playing to broader inter-Arab themes.[27]

Israel is understandably averse to seeing its nuclear advantage eroded. Indeed, as demonstrated by its 1981 strike on the Iraq Osirak reactor, Israel is ready to maintain a nuclear monopoly in the region through the use of force. That the world was a safer place following the destruction of the Iraqi reactor is undoubted. And it is noteworthy that Iraq possessed a relatively advanced military program despite its membership in the NPT. However, in the long run, the tactical success of the air raid may prove to be counterproductive. It certainly underscored the Arab belief that their defences could only be marginally effective against Israel in the absence of a credible nuclear deterrent. Moreover, Arab governments reacted with mixed emotions of indignation and suspicion at Israel's de facto claim to a monopoly. Applying the logic of the East-West balance of power persuaded some in the region of the stability that nuclear proliferation might impart. This logic may be dubious, but it nevertheless provides a powerful incentive for countries seeking a nuclear weapons capability.[28]

These dynamics tend to reinforce themselves. Iraq's weapons programmes may have been intended to confront Israel, but it was the Irani-

---

[27] Donovan, Michael, 'Iran, Israel and Nuclear Weapons in the Middle East,' 14 February 2002, p. 3.

[28] Ibid, pp. 3–4.

ans who suffered from Saddam's obsession with chemical weapons as the world, including the US, observed in silence during their eight-year war. This instilled in Tehran the powerful lesson that it must be responsible for its own defence. US acquiescence to the Israeli nuclear programme further erodes Tehran's faith in the equal application of international arms regimes. They hesitate to ratify the Chemical Weapons Convention (CWC) despite their own doubts as to the utility of these weapons. Iranian attempts to hedge against perceived threats couched as their incendiary rhetoric aimed at Israel, simply reinforces Iran's image as a proliferator.[29]

There was a slight shift in the Israeli position in this regard. The Israelis still refuse to deal with the matter within international parameters. However, they have agreed to start negotiations to establish an NWFZ in the Middle East in case of substantial progress in the peace process with their neighbours. This stance varies to some extent from Israel's previous position as it was not ready to engage in any negotiations relating to arms control unless peace was established in the region.

Iran has proposed the creation of a WMD-free zone in the Middle East, but this approach is disingenuous. The idea links Israel's nuclear weapons capability to an *eventual* political settlement between the Arabs and Israel that Iran actively opposes. Such a strategy seeks to place the burden of proof on Israel's intention to forgo its nuclear deterrent. Tel Aviv will not address the nuclear question without a *prior*, far-reaching political settlement and several years of confidence-building measures. Thus the mechanics for a regional nuclear arms race are, for the time being, locked into place. The Israeli nuclear arsenal will continue to drive other States in the region to WMD acquisition efforts for the foreseeable future. In turn, these ambitions are likely to underscore Israel's sense of vulnerability. Strategies of containment and technology denial may slow the process, but they offer no permanent solutions.[30]

Successive Israeli Governments have expressed their worries about moves towards the acquisition of nuclear weapons by Arab countries on one hand and, by Iran, in particular, on the other. According to Cohen and Miller, "Rabin, when he was the Prime Minister of Israel, believed that the Gulf War had given Israel a window of opportunity of perhaps 5 or 10 years to minimise the threat of Israel's enemies of getting hold of nuclear weapons. They consider that during this period, Israel should

---

[29] Ibid, p. 4.
[30] Ibid.

contribute to the vigorous nuclear denial strategy via enhanced political and intelligence coordination with friendly States and, more fundamentally, it should negotiate peace agreements with its neighbours to reduce incentives and support for nuclearisation in the Arab World."[31] In the meantime, they suggest that Israel might agree to limit its nuclear capability as a step towards an NWFZ in the Middle East and the "Arab States and Iran should reinforce their declaratory authority of the IAEA to make special inspections at both declared and suspect nuclear facilities." This proposal was incorporated into the Bush initiative on arms control in the Middle East that was put forward in a speech at the Air Force Academy in Colorado on 29 May 1991, shortly after the Gulf War.[32]

The reactions of Israel and the Arabs to the proposal were not favourable. The Arabs are unwilling to accept, even as an interim measure, an Israeli monopoly of nuclear weapons. However, this *is* an existing fact. On the other hand, Israel is unwilling to accept safeguards inspections on the Dimona Nuclear Centre. Moreover, the Israeli leadership is unwilling to have an open debate about nuclear weapons. Nevertheless, Cohen and Miller suggest that "Israel should shut down the Dimona reactor, and hence cease its production of plutonium, as a unilateral undertaking and convey assurances about its nuclear doctrine in private discussions and through third parties, especially the US."[33] This idea may prove to be a realistic one. Israel may be persuaded to shut down Dimona soon, being an aging 30 year old reactor. Moreover, Israel has already produced far more plutonium than it needs for any rational purpose. This could be the sole step Israel might assume until the peace process succeeds and Israel reaches the satisfaction that the risks of further nuclear weapon proliferation in the Middle East are minimal. Even then, the Israeli political leadership will have to counter those who argue that nuclear weapons are even more necessary to guarantee the security of a smaller Israel. Of all the world's nuclear weapons, Israel's will prove the most difficult to get rid of.[34]

When the political and military conditions in the region are sufficiently adjusted to allow for the resumption of the diplomatic process,

---

[31] Barnaby, Frank, 'Capping Israel's Nuclear Volcano,' in Efraim Karsh (ed.), 'Between War and Peace: Dilemmas of Israeli Security,' 1996, p. 110.

[32] Ibid, pp. 109–110.

[33] Ibid, p. 110.

[34] Ibid, p. 111.

regional and co-operative security frameworks will be on the agenda again. In order to make progress, it will be necessary to avoid repeating the mistakes of the previous round.[35]

Based on the experience and lessons of the 1990s, the following is required:[36]

- A major investment of political resources towards the transformation of political and cultural perceptions, and the development of relations based on mutual recognition and acceptance;
- Acknowledgment of fundamental asymmetries, their consequences in terms of national security requirements, and the role of mutual verification mechanisms;
- Step-by-step development beginning with confidence and security building measures, and including cooperation on issues such as small arms, landmines and conventional weapons;
- Consistent, responsible and effective action by the international community to insure that signatories honour their commitments with respect to treaties such as the NPT, CWC and the Biological and Toxin Weapons Convention; and
- Effective control over the export of technologies and expertise used in the proliferation of WMD and missile capabilities in the region.

These steps can be reasonably achieved if Israel does not bind them to the peace process. The idea of transforming political and cultural perceptions needs more than decades to be achieved. This even goes beyond the Israeli demand to bind the establishment of an NWFZ in the region to calls for achieving peace or at least achieving major developments in the peace process.

At the beginning of the 1990s, following the end of the Cold War, the prospects for developing a foundation for regional security, and for preventing the proliferation of WMD and ballistic missiles in the Middle East, appeared to be promising. This optimism led to major investments in time and resources designed to promote the development of regional security and arms control frameworks. However, the stagnant political relationships in the region, the difficulties associated with the asymmetries, the failure to construct a foundation for cooperation based on confidence-building measures, and the increasing threat posed by

---

[35] Steinberg, Gerald M., 'Starting over: the prospects for regional security and arms control in the Middle East in the next decade,' Disarmament Forum, Issue 2, 2001, p. 74.

[36] Ibid.

WMD in Iraq, as well as other countries in the region combined to produce meagre results. If this outcome is to change when the next window of opportunity opens, the lessons from the first round should be understood.[37]

Events in the Middle East proved that this optimism is premature and, at the end of the 1990s, the potential for regional agreements and limitations in the Middle East seem very limited, at least in the immediate future. A combination of factors explain this disappointing outcome, including: 1) the failure to create the political and security environment to allow for the development of cooperative security; 2) asymmetry and the limits of universality; 3) the absence of confidence-building measures; 4) the inconsistent implementation of the terms and undertakings with respect to disarming Iraq; and 5) the continued flow of WMD and missile technology in the region, thereby heightening threat perceptions. Each one of these deficiencies, in itself, was very costly to the process, and the combination proved fatal. It led to a lengthy futile process with limited progress in the Arms Control and Regional Security.[38]

States in the region still lack confidence. Constant hidden Israeli threats of its being ready to use nuclear weapons against any State in the region if attacked by its neighbours and the unhealthy political environment in the Middle East, delay any measures that could be undertaken to build mutual confidence.

The Iraqi WMD threats are over, and so far, no State in the Middle East, other than Israel, produces or possesses nuclear weapons. The international community acts very strictly with any State in the region that may jeopardise the NPT. That was corroborated in the cases of Iraq, Libya and Iran. However, the prolongation of the flow of WMD and missile technology in the region still intensifies the threats of proliferation. However, when it comes to nuclear weapons alone, time and recent developments confirm that Israel is trying to maintain its nuclear monopoly in the region, while still maintaining its official position that it will not be the first to introduce nuclear weapons into the region. It is a false pretence to claim that Israel is not the only State that possesses nuclear weapons in the Middle East.

Pending the improvement of circumstances in the region, efforts to force the pace and skip stages are likely to be counterproductive. Without first building the infrastructure to support a credible regional security

---

[37] Ibid, p. 1.

[38] Ibid, p. 71.

framework, and creating a foundation of stability and security, States facing major threats, including Israel, will not be willing to dismantle their basic defence and deterrence structures. Pressures to force premature disarmament, in the absence of reliable alternative security structures, will be seen as efforts to gain unilateral advantage in a zero-sum context, rather than part of a broader approach towards co-operative security and stability. The road to regional security must be travelled one careful step at a time.[39]

The dilemma facing the region is that the political circumstances which paved the way for Iraq to pursue its clandestine activities may rise again with other countries in the region or outside. Developing the system of the NPT and the safeguards are important to give a real incentive to States, especially Israel, to accept negotiations on nuclear disarmament in the Middle East. If there are enough security assurances, not only from the Arab States and Iran, but also from NWS (by implementing Article VI of the NPT), Israel may consider altering its stance towards establishing an NWFZ in the region.

From a legal perspective, if Israel ratifies the NPT and places its nuclear facilities under IAEA safeguards, the region will be free from the threat of nuclear weapons. Due to political realities in the region, the Israeli deterrent is not effective with respect to the balance of power in the region. Israel would be affected negatively by using a nuclear weapon even more than its Arab adversaries, especially if they have chemical and biological weapons (as Israel claims).

In addition, Israel is obligated as a UN member to implement the resolutions of the UN organs. Hence, Israel is obligated under SC resolution 487 (1981) and all the GA resolutions since 1974 to adhere to the NPT and place its nuclear facilities under IAEA safeguards. It is also under the obligation to undertake all necessary measures to assist in establishing an NWFZ in the Middle East.

The fact that the suggestion to establish an NWFZ is still alive and capturing the attention of the world provides hope that one day we may wake up to find a Middle East free from nuclear weapons, and hopefully from all WMD.

---

[39] Ibid, p. 76.

CONCLUSIONS

There is not the least doubt that nuclear technology has been one of the most controversial issues in recent times. The nature of its dual-usage became the epicentre of an ongoing debate where rules were established to encourage its peaceful uses and prohibit its military ones. The assessment of this regime leads to the following conclusions:

1. The rules of the NPT and those of the IAEA governing the transfer of nuclear technology proved to be more restrictive than what the NNWS originally envisaged. The book concludes that this is mainly due to the absence of a procedural system to ensure that NNWS exercise their inalienable right to gain access to nuclear science, material, equipment and know-how. The book went further to conclude that the implementation of the NPT/IAEA rules inhibits the exercise of such a right.
2. The attempt to strike a balance between the monitoring role of the IAEA and State sovereignty reveals the inherent contradiction between international law and State sovereignty. The current developments in the IAEA safeguards system, the expansion of inspection rights, and the adoption of the Model Additional Protocol to the Safeguards Agreements (INFCIRC/540), constitute a regression in the concept of State sovereignty, the main pillar upon which international law is formulated.
3. There is a loophole in the NPT and IAEA rules governing nuclear supplies. The restrictions imposed on NNWS to use nuclear technology (material, equipment and know-how), are only directed to the Recipient or Beneficiary States rather than Supplier States. The latter are not obligated to report their nuclear supplies to the Agency. Hence, the IAEA highly depends on the cooperation of one party (the recipient) in a supply agreement, which negatively affects the implementation of the safeguards system.
4. The different patterns in the international community's reactions towards States attempting to build nuclear weapons or any other explosive device and the absence of a coherent norm in these

reactions obstructs the efforts to achieve the universality of the NPT.

5. The Supplier Groups, established to control the movement of nuclear materials around the world, impose supply criteria that go beyond the NPT requirements. These criteria are more subjective than objective. Hence, they negatively affect the transfer of nuclear technology for peaceful purposes. This constitutes a violation by the Supplier States of their obligations under the NPT.
6. The IAEA is an executive organ monitoring the non-diversion of nuclear technology from peaceful to military uses. However, it simultaneously plays the role of a party and a judge while deciding cases of non-compliance. In such cases, the IAEA refers the matter to the SC. As a political organ, the SC may adopt punitive measures against the non-compliant State. Hence, the SC becomes a judge as well. In light of the absence of an independent legal organ to judge on such matters and to avoid the impartiality of the NPT/IAEA regime, the book suggests that this role could be performed by an ad hoc committee (for example, an arbitration panel) or the ICJ.
7. The NPT rules are inadequate to enforce Article VI. It does not set out a time frame for the commencement of negotiations among NWS to cease the nuclear arms race or to reduce their nuclear arsenals as a primary step towards a complete elimination of nuclear weapons.
8. The book suggests that one of the reasons behind the non-universality of the NPT is the reliance of NWS on nuclear weapons as the strategic weapon in their deterrence policies.
9. The Iraqi case was the turning point that exposed failures in the safeguards system as outlined in the book. In an attempt to rectify such failures, the IAEA adopted the Model Additional Protocol. However, the book concludes that although the protocol has strengthened the current safeguards system to a large extent, nevertheless, it does not grant the IAEA similar inspection rights to what the Agency enjoys under the SC resolutions against Iraq.
10. The previously stated discrepancies in the international legal system seem to be among the reasons behind the delay in establishing an NWFZ in the Middle East. The book proposes that these discrepancies along with the existence of nuclear weapons and the shortages in the safeguards system add complexities to an already volatile region.
11. The Israeli case is not easy to handle. Israel continues to ignore the implementation of the SC resolution 487 and the annual GA resolutions ever since the Israeli nuclear armament was addressed

back in 1974. Consequently, the book concludes that Israel is in violation of its international obligations under Articles 2 and 25 of the UN Charter as long as it retains its nuclear arsenal.

12. The book suggests that the decisions by the BG requesting Iran to halt all its enrichment activities and not to construct a nuclear heavy water power plant as a confidence-building measure constitute a breach of Article IV of the NPT. If the ongoing attempts to indefinitely prohibit Iran from resuming its nuclear activities in the aforementioned fields succeed, this shall constitute a precedent to further limiting NNWS rights to acquire nuclear technology for peaceful purposes.
13. In order to reach its goals and to bridge the inherent gaps in the rules, the NPT must be amended. The book proposes the adoption of a protocol to amend Articles III, IV and VI of the NPT to provide for the following:
    - Assuring the access of NNWS to nuclear material and technology, by initiating a transparent system of cooperation between supplier and recipient States.
    - Supplier States should inform the IAEA of their nuclear transfers to NNWS in an orderly and a timely manner to facilitate the implementation of the safeguards system.
    - Adopting efficient measures to ensure that NWS implement their obligations under Article VI of the Treaty, through setting out a time table to halt the production of fissile materials, to cease manufacturing and developing new nuclear explosive devices, and to negotiate concrete steps towards reducing the existing nuclear weapons in order to reach the complete elimination of nuclear weapons.

Until the political situation in the Middle East is conducive to negotiate the establishment of an NWFZ, the book concludes that the amendment of the NPT rules should instigate a positive effect on the region. Hence, the dream to live in a region free from nuclear weapons may one day become true and cease from being a mere mirage.

BIBLIOGRAPHY

Abd El Salam, Mohamed, 'Nuclear Weapon Free Zones between the Theoretical Conditions and the Practical Experiences,' Al-Ahram Centre for Political and Strategic Studies, Al-Ahram Publications, Cairo, 2003. (*Arabic Language*)

Abd El Salam, Mohamed, 'The Limits of Power; the Uses of the Israeli Nuclear Weapons,' Al-Ahram Centre for Political and Strategic Studies, Al-Ahram Publications, Cairo, 1996. (*Arabic Language*)

Acharya, Amita V. and Sola Ogunbanwo, 'The Nuclear Weapon-Free Zones in South-East Asia and Africa,' SIPRI Yearbook, Oxford University Press, UK, 1998.

AFRA homepage, http://www.afra-iaea.org/template1.asp?catId=5&parentId=0 &childId=0& levelId=0&node=1#

Agrawala, S.K., 'The Role of the General Assembly Resolutions as Trend-Setters of State Practice,' The Indian Journal of International Law, Volume 21, Indian Society of International Law, New Delhi, 1981.

Albright, David and Corey Hinderstein, 'Iran: Furor over Fuel,' Bulletin of Atomic Scientists, Volume 59, No. 3, Chicago, May/June 2003, http://www.thebulletin.org/issues/2003/mj03/mj03albright.

Albright, David and Mark Hibbs, 'Iraq's shop-till-you-drop nuclear program,' Bulletin of Atomic Scientist, Volume 42 no. 3, Chicago, April 1992. http://www.thebulletin.org/article.php?art_ofn=apr92albright.

Albright, David and Khidhir Hamza, 'Iraq's Reconstitution of Its Nuclear Weapons Program,' Arms Control Today, the Monthly publication of Arms Control Association, Washington, DC, October 1998, http://www.isis-online.org/publications/iraq/act1298.html.

Almond, Jr., Harry H., 'Nuclear Weapons, Nuclear Strategy and Law,' Denver Journal of International Law and Policy, Volume 15, Number 1, University of Denver College of Law, Denver, Colorado, summer 1986.

Alves, Péricles Gasparini and Daiana Belinda Cipollone, 'Nuclear-Weapon-Free Zones In The 21st Century,' United Nations Institute for Disarmament Research, Geneva, United Nations publication, New York, June 1997.

The American Assembly, 'Atoms for Power: United States policy in Atomic Energy Development,' Columbia University, New York, December 1957.

US Department of Defence, 'Nuclear Posture Review 2001,' Annual Defence Department Report, Report to the President and Congress, submitted to the Congress on 31 December 2001. www.globalsecurity.org/wmd/policy/dod/npr.htm.

Baer, Alec, 'Nuclear Supplier Group and its Time,' paper presented before the 2nd Seminar on Export Controls in Nuclear Non-Proliferation held at UN in New York, 8 to 9 April 1999.

Barnaby, Frank, 'Capping Israel's Nuclear Volcano,' in Efraim Karsh (ed.), 'Between War and Peace: Dilemmas of Israeli Security,' Frank Cass & Co. Ltd., London, 1996.

Barnard, Chester I., J.R. Thomas, Charles A. Oppenheimer, Harry A. Winne, and David E. Lilienthal, 'A Report on The International Control of Atomic Energy,' Department of State, Washington, D.C, March 1946.

Baute, Jacques, Director of Iraq Nuclear Verification Office, in an interview held at the IAEA Headquarters in Vienna, May 2004. (Meeting)

Beres, Louis Rene, 'Israel's Bomb in the Basement: A Revisiting of "Deliberate Ambiguity" vs. "Disclosure," in Efraim Karsh (ed.), 'Between War and Peace: Dilemmas of Israeli Security,' Frank Cass & Co. Ltd., London, 1996.

Bertel, Eveline and Joop van de Vate, 'Nuclear Energy and the Environmental Debate: the Context of choices,' through international bodies on climate change, the roles of nuclear power and other energy options are being assessed, International Atomic Energy Bulletin, Volume 37, issue 4, Vienna, 1995, http://www.iaea.org/Publications/Magazines/Bulletin/Bull374/37404080207.pdf.

Betts, Richard K., 'The New Threat of Mass Destruction,' Foreign Affairs, Volume 77, No. 1, 1998.

Bhatia, Shyam and Daniel McGrory, 'Brighter then Baghdad Sun, Saddam Hussein's Nuclear Threat to the United States,' Regency Publishing, Inc, Washington, 2000.

Biad, Abdelwahab, 'Nuclear-Weapon-Free Zones: Challenges and Prospects for Non-Proliferation,' Nuclear Law Bulletin, Issue 58, Nuclear Energy Agency, OECD, Paris, 1996.

Blackett, P.M.S., 'Military and Political Consequences of Atomic Energy,' Turnstile Press Ltd., London, 1948.

Blood, Peter R., Foreign Affairs, Defense and Trade Division, 'Pakistan-US Relations,' CRS Issue Brief for Congress received through the CRS web, updated 10 March 2002, Congressional Research Service. The Library of Congress, Order Code IB94041, http://fpc.state.gov/documents/organization/9057.pdf.

Blix, Hans, 'The Peaceful and Safe Uses of Nuclear Energy,' paper in the IAEA/OPNAL Seminar on 'IAEA Safeguards: Verifying Compliance with Non-Proliferation Commitments,' Jamaica, April 1996.

Blix, Hans, 'Disarming Iraq,' Pantheon Books, New York, 2004.

Bogdanov, O., 'Détente and Disarmament the Legal Aspects,' Review of Contemporary Law, Brussels, 1977.

Boustany, Katia, 'Reflection on the Development of Nuclear Law,' Legal Sciences Department University of Quebec, Montreal, Nuclear Law Bulletin, Bulletin 51, Nuclear Energy Agency, OECD, Paris, June 1993.

Boustany, Katia, 'The Development of Nuclear Law-Making or the *Art of Legal "evasion"*,' Nuclear Law Bulletin, Bulletin 61, Nuclear Energy Agency, OECD, Paris, June 1998.

Bunn, George, 'Inspection Clandestine Nuclear Activities: Does the Nuclear Non-Proliferation Treaty Provide Legal Authority for the International Atomic Energy Agency's Proposals for Reform?' Nuclear Law Bulletin, Bulletin 57, Nuclear Energy Agency, OECD, Paris, June 1996.

Bunn, G., R.M. Timerbaev and J.F. Leonard, 'Nuclear Disarmament: How Much Have the Five Nuclear Powers Promised in the Non-proliferation Treaty?' the Lawyers Alliance for World Security, the Committee for National Security and the Washington Council on Non-Proliferation, June 1994.

Burn, D., 'The Political Economy of Nuclear Energy,' Institute for Economic Affairs, London, 1967.

Canterbery, E.R., B. Johnson and D. Reading, 'Cost Savings from Nuclear Regulatory Reform: An Econometric Model,' Southern Economic Journal, Volume 62, no. 3, 1996.

Cassidy, Stephen H., 'The Newest Member of the Nuclear Club: Pakistan's Drive for a Nuclear Weapons Capability and United States Nuclear Non-proliferation Policy,' Hastings International Comparative Law Review, Volume 12, No. 1, University of California, fall 1988/1989.

Castañeda, Jorge, 'Legal Effects of United Nations Resolutions,' Number 6, Columbia University Studies in International Organisation, Columbia University Press, New York and London, 1969.

Catudal, Jr., Honore M., 'Israel's Nuclear Weaponry—A New Arms Race in the Middle East,' Grey Seal Books, London, 1991.

Cohen, Avner, 'Beyond the Pale: Israel and the Bomb,' Bulletin of Atomic Scientists, Volume 55, Issue 1, Chicago, January/February 1999.

Cohen, Avner, 'Israel and the Bomb,' Columbia University Press, New York, 1998.

CERN, http://public.web.cern.ch/Public/Content/Chapters/AboutCERN/WhatIsCERN/WhatIsCERN-en.html

'Convention for the Establishment of a European Organisation for Nuclear Research,' Paris, 1 July 1953, amended in 17 January 1971. http://legal-service.web.cern.ch/legal-service/conventionE.htm.

Cordesman, Anthony H., 'Weapons of mass destruction in the Middle East,' Brassey's, London, 1991.

Committee on Atomic Energy of the Carnegie Endowment for International Peace, 'Utilisation and Control of Atomic Energy,' A Draft Convention prepared by the Legal Subcommittee in consultation with other Legal, Political and Scientific Experts, New York, June 1946.

Crawford, James and Phillipe Sands, 'Legal Aspects of Nuclear Weapons Convention,' African Yearbook of International Law, Volume 6, Kluwer Law International, 1998.

Daneke, Gregory A., 'Energy, Economics and the Environment: Towards a Comprehensive Perspective,' Lexington Books, Lexington, Mass., 1982.

De La Fuente, A. Herrero, 'The Legal Force of International Rules Relating to Nuclear Risks,' Nuclear Law Bulletin, Bulletin 30, Nuclear Energy Agency, OECD, Paris, December 1982.

Dean, Jonathan, 'Expanding the Security Council Role in Blocking the Spread of Nuclear Weapons,' Transnational Law & Contemporary Problems, Volume 2, Number 1, A Journal of University of Iowa College of Law, Spring 1992.

Deaver, Micheal A., 'Disarming Iraq, Monitoring Power and Resistance,' Praeger Publishers, Westport, Conn., 2001.

Dimitrieva, Galina Kirillovna and, Igor Ivanovic Lukashuk, 'The role of the UN General Assembly Resolutions in the International Norm-Making,' The Indian Journal of International Law, Volume 28, Issue 2, Indian Society of International Law, New Delhi, 1988.

Donovan, Michael, 'Iran, Israel and Nuclear Weapons in the Middle East,' Terrorism Project of the Centre for Defence Information, Washington D.C., 14 February 2002.

Doub, William O. and Eugene R. Fidell, 'International Relations and Nuclear Commerce: Developments in United States Policy,' Law and Policy in International Business, Volume 8, Number 4, Georgetown University Law Center, Washington, DC, 1976.

DPA German Press Agency, 'Olmert Appears to Admit Israel Possesses Nuclear Weapons,' 11 December 2006.

Ellis, Mark E., 'The New International Economic Order and General Assembly Resolutions: The Debate over the Legal Effects of General Assembly Resolutions Revisited,' California Western International Law Journal, Volume 15, US International University, California Western School of Law, San Diego, CA, 1985.

Epstein, William and Paul Szasz, 'Giving the Non-Proliferation Treaty Teeth: Strengthening the Special Inspection Procedures of the International Atomic Energy Agency,' Virginia Journal of International Law, Virginia, summer 1993.

Epstein, William, 'NPT Wrap- up, Indefinite Extension with Increased Accountability,' Bulletin of the Atomic Scientists, Volume 51, No. 4, Chicago, July/August 1995. http://www.the bulletin.org/index.htm.

European Union homepage, http://europa.eu.int/comm/euratom/index_en.

Evans, Senator, 'Statement on 18 April 1995 to the 1995 Review and Extension Conference of the States Parties to the Treaty on Non-Proliferation of Nuclear Weapons,' 1995, http://www.dfat.gov.au/isecurity/pd/pd_sept_95/peace5.html.

Federal Ministry for European and International Affairs and the Institute for International Law and Justice at New York University School of Law, 'The Role of the Security Council in Strengthening a Rules-based International System,' Final Report and Recommendations from the Austrian Initiative, 2004–2008, http://www.bmeia.gv.at/fileadmin/userupload/bmeia/media/Vertretungsbehoerden/New_York/Kandidatur_SR/FINAL_Report_The_UN_Security_Council_and_the_Rule_of_Law.pdf

Feldman, Shai, 'Israeli Nuclear Deterrence, A Strategy for the 1980s,' Columbia University Press, New York, 1982.

Feldman, Shai, 'Nuclear Weapons and Arms Control in the Middle East,' (BCSIA Studies in International Security), Centre for Science and International Affairs, John F. Kennedy School of Government Harvard University, Cambridge Massachusetts, 1997.

Fischer, David, 'History of the IAEA, the First Forty Years,' published by IAEA, Vienna, September 1997.

Forlati, L. and A. Giardina, 'Disarmament and the United Nations,' International Association of Democratic Lawyers, Review of Contemporary Law, Brussels, 1977.

Ford Foundation, 'A Time to Choose: America's Energy Future; Final Report to the Energy Policy Project of the Ford Foundation,' Ballinger Publication, Cambridge, Mass, 1974.

Freeman, Dyson, Scientific American 150th Anniversary Issue: 'Key Technologies for the 21st Century,' W.H. Freeman, New York, 1996.

Friends of the Earth Europe, 'Abolition of the EURATOM Treaty,' 2002, http://www.foeeurope.org/activities/Nuclear/pdf/2002/Euratom_briefing.pdf.

Gangl, Walter T., 'The *jus cogens* Dimensions of Nuclear Technology,' Cornell International Law Journal, Volume 13, Number 1, Cornell University, New York, Winter 1980.

GA, Resolution 3263 (XXIX), 'Establishment of Nuclear-Weapon-Free Zone in the Region of the Middle East,' 9 December 1974.

GA, A/36/431, Report of the SG of the UN, 'Israel Nuclear Armament,' Distribution General, 18 September 1981.

GA, Resolution A/RES/36/98, 'Israeli Nuclear Armament,' Distribution General, 9 December 1981.

GA, A/40/520, Report of UNIDIR, 'Israeli Nuclear Armament,' Distribution General, 9 August 1985.

GA, Resolution A/RES/95/25, 'Sustainable Fisheries, including through the 1995 Agreement for the Implementation of the Provisions of the United Nations Convention on the Law of the Sea of 10 December 1982 relating to the Conservation and Management of Straddling Fish Stocks and Highly Migratory Fish Stocks, and related instruments,' Distribution General, 17 January 2005.

GA, SC, A/61/735–S/2007/73, 'Letter dated 8 February 2007 from the Permanent Representative of Cuba to the United Nations addressed to the Secretary-General,' 9 February 2007.

Gallucci, Robert, Former Deputy Executive Director of UNSCOM, 'Reflections on Establishing and Implementing the Post-Gulf War Inspections of Iraq's Weapons of Mass Destruction Programs,' a speech given in a seminar 'Understanding the Lessons of Nuclear Inspections and Monitoring in Iraq: A Ten-Year Rreview,' sponsored by the Institute for Science and International Security, 14–15 June 2001. Transcript date: 31 July 2001, http://www.isis-online.org/publications/Iraq/Gallucci

Gilpin, Robert, 'American Scientists and Nuclear Weapons Policy,' Princeton University Press, Princeton, 1962.

Goldblat, Jozef, 'Twenty Years of the Non-Proliferation Treaty-Implementation and Prospects,' International Peace Research Institute, Oslo, 1990.

Goldblat, Jozef, 'The Non-Proliferation Treaty: How to Remove the Residual Threats,' United Nations Institute for Disarmament Research, Research Paper No. 13, United Nations Publications, New York, 1992.

Goldblat, Jozef and Ragnhild Ferm, 'UN General Assembly Resolutions and Decisions on Arms Control and Disarmament 1985–1986,' SIPRI Yearbook, Oxford University Press, 1987.

Goldenberg, Suzanne, 'Pakistan Helped North Korea Make Bomb,' the Guardian, 19 October 2002.

Government of India, Policy Planning and Research Division—Ministry of

External Affairs, 'India—Bilateral Treaties and Agreements,' SIBA EXIM PVT.LTD, Volume 1:1947–1952, Delhi, 1994.

Hallahmi, Benjami Beit, 'The Israeli Connection—Whom Israel Arms and Why?' I.B. Tauris & Co. Ltd. Publishers, London, 1987.

Hammad, Fawzy, 'The Role of the Atomic Energy Authority in the Peaceful Uses of Nuclear Energy,' Atomic Energy Authority Publications, Cairo, 1991. (*Arabic Language*)

Hammad, Fawzy, Dr. and Adel Mohamed Ahmed, 'Problems of Establishing a Nuclear Weapon Free Zone in the Middle East,' El- Seyasa El-Dawlya (International Politics), Issue 145, October 2001, Al-Ahram Centre for Political and Strategic Studies, Cairo, 2001. (*Arabic Language*)

Hammad, Fawzy H. and Adel M. Ali, 'Principles of Establishing a Middle East Weapons of Mass Destruction Free Zone Monitoring and Verification System,' published in 'Building a Weapons of Mass Destruction Free Zone in the Middle East: Global Non-Proliferation Regimes and Regional Experiences,' UNIDIR and League of Arab Nations, Geneva, December 2004.

Hammad, Fawzy, Former Chairperson of the Egyptian Atomic Energy Authority, in a Meeting held at Cairo, in November 2003. (Meeting)

Hamza, Khidhir 'Inside Saddam's secret nuclear program,' Bulletin of the Atomic Scinetists, Volume 54, No. 5, September/October 1998.

Hamza, Khidir with Jeff Stein, 'Saddam's Bombmaker, The Terrifying Inside Story of the Iraqi Nuclear and Biological Weapons Agenda,' Scribner, New York, 2000.

Hanford Health Information Network, 'A Monograph Study of the Health Effects of Radiation and Information Concerning Radioactive Releases from the Hanford Site: 1944–1972,' September 1996. http://www.doh.wa.gov/hanford/publications/health/rad-home.htm

Harris, Robert, 'Alternative Energy Resources: An International Approach,' published in 'Regulation of Atomic Energy,' Columbia Journal of Transitional Law, Volume 16, 1977.

Harkavy, Robert E., 'Spectre of A Middle Eastern Holocaust: The Strategic and Diplomatic Implications of the Israeli Nuclear Weapons Programme,' University of Denver for Monograph Series in World Affairs, Volume Fourteen, Book Four, Denver, Co., 1977.

Higashi, J., 'The Role of Resolutions of the United Nations General Assembly in the Formative Process of International Customary Law,' The Japanese Annual of International Law, Volume 25, The International Law Association of Japan, Tokyo, 1982.

Hitchens, Theresa, Vice President, Centre for Defense Information, 'Slipping Down the Nuclear Slope: Bush Administration Nuclear Policy Lowers Bar Against Usage,' Presented at a conference on 'US Nuclear Policy and Counter Proliferation,' 26 February, 2003, http://www.cdi.org/nuclear/counterproliferation-conference.cfm.

Horbach, Nathalie L.J.T., 'Assistance Programmes of the International Atomic Energy Agency to CEEC/NIS,' published by the Section on Energy and Natural Resources Law of the International Bar Association, in association with LLP Ltd. University of Dundee, Journal of Energy and Natural Resources Law,

Volume 17, No. 3, August 1999.

IAEA, 'Nuclear Law for a Developing World'—Lectures given at a Training Course—Vienna 16–26 April 1968, IAEA Legal Series No. 5, Vienna, 1969.

IAEA, 'Bilateral, Regional and Multilateral Agreements Relating to Cooperation in the Field of Nuclear Safety,' IAEA Legal Series No. 5, Vienna, 1990.

IAEA, 'Joint FAO/IAEA Programme, Nuclear Techniques in Food and Agriculture —Supporting Research for Agricultural Development,' Sub-Programme on 'Animal Production and Health,' www.iaea.org/programmes/nafa/d3/d3.html.

IAEA, 'FAO/IAEA Training and Reference Centre—General Information,' www.iaea.org/trc/gen-info-trc.html.

IAEA, 'Record of the Seventh Plenary Meeting for the General Conference,' held at Vienna, September 1995.

IAEA, 'International Atomic Energy Agency: Personal Reflections,' Vienna, 1997.

IAEA, [GC (44)/9], 'Nuclear Technology Review 2000,' Forty-Fourth Regular Session of the General Conference of the Agency, Vienna, 14 August 2000.

IAEA, 'Nuclear Power in An Era of Change,' Statement by the Mr, El Baradei, DG of IAEA to the Pacific Basin Nuclear Conference, Seoul, Republic of Korea, 2 November 2000, https://www.iaea.org/NewsCenter/Statements/2000/ebsp2000n012.shtml.

IAEA, INFCIRC/66/Rev.2, 'The Agency's Safeguards System (1965, as Provisionally Extended in 1966 and 1968),' General Distribution, 16 September 1968.

IAEA, GOV/2530, Board of Governors, 'A Report by the DG on Non-Compliance by Iraq with its Obligations under the Safeguards Agreement Concluded with the Agency,' Restricted Distribution, 16 July 1991.

IAEA, 'The Technical Cooperation Programme,' http://www-tc.iaea.org/tcweb/tcprogramme/default.asp.

IAEA, GOV/INF/824, Board of Governors, 'Technical Cooperation Strategy,' 24 October 1997. http://www-tc.iaea.org/tcweb/tcprogramme/ginf824.pdf

IAEA, INFCIRC/153, 'The Structure and Content of Agreements between the Agency and States Required in Connection with the Treaty on the Non-Proliferation of Nuclear Weapons,' Distribution General, June 1972, http://www.iaea.org/Publications/Documents/Infcircs/Others/inf153.shtml.

IAEA, INFCIRC/377, 'African Regional Co-operative Agreement for Research, Development and Training Related to Nuclear Science and Technology,' Vienna, 2 April 1990.

IAEA, INFCIRC/540, 'Model Protocol Additional to the Agreement(s) between State(s) and the International Atomic Energy Agency for the Application of Safeguards,' Printed by the IAEA in Austria, September 1997. http://www.iaea.or.at/Publications/Documents/Infcircs/1997/infcirc540.pdf# search= 'infcirc/540'

IAEA, INFCIRC/539/Rev.2, 'Communication of 28 August 2003 received from the Government of the US on behalf of the Member States of the Nuclear Suppliers Group,' 16 September 2003.

IAEA, INFCIRC/254/Rev.6/Part 1, 'Communication Received from Certain Member States Regarding Guidelines for the Export of Nuclear Material, Equipment and Technology,' 16 May 2003.

IAEA, INFCIRC/254/Rev.5/Part 2, 'Communication Received from Certain Member States Regarding Guidelines for Transfers of Nuclear Related Dual- Use Equipment, Materials, Software and Related Technology,' 16 May 2003.

IAEA, INFCIRC/637, 'Communication dated 26 November 2004 received from the Permanent Representative of France, Germany, the Islamic Republic of Iran and the United Kingdom,' concerning the Agreement Signed in Paris on 15 November 2004.

IAEA, INFCIRC/723, 'Communication dated 4 March from the Governor for the Russian Federation and the Resident Representatives of China, France, Germany, the United Kingdom and the United States of America concerning UN SC resolution 1803 (2008),' 5 March 2008.

IAEA, INFCIRC/724, 'Communication dated 26 March 2008 received from the Permanent Mission of the Islamic Republic of Iran to the Agency,' 28 March 2008.

IAEA, INFCIRC/711, 'Communication dated 27 August 2007 from the Permanent Mission of the Islamic Republic of Iran to the Agency Concerning the Text of the "Understandings of the Islamic Republic of Iran and the IAEA on the Modalities of Resolution of the Outstanding Issues,"' 27 August 2007.

IAEA, INFCIRC/729, 'Communication dated 16 June 2008 received from the Permanent Mission of the Islamic Republic of Iran to the Agency Concerning the Text of the Islamic Republic of Iran's Proposed Package for Constructive Negotiation,' 18 June 2008.

IAEA, INFCIRC/750, 'Communication dated 25 June 2008 received from the Resident Representative of the United Kingdom to the Agency Concerning a Letter and Offer of 12 June 2008 Delivered to the Islamic Republic of Iran,' 1 July 2008.

IAEA, 'Transcript of the DG's Press Statement on IAEA Inspection in Iran,' IAEA Headquarters, Vienna, 18 June 2004.

IAEA, S/25621, 'Note by the SG,' 19 April 1993, annexed to it a 'Letter dated 8 April 1993 from the Director General of the IAEA addressed to the SG,' attached to it the 'Third Report of the DG of the IAEA on the Implementation of the Agency's Plan for Future Ongoing Monitoring and Verification of Iraq's Compliance with Paragraph 12 of Resolution 687 (1991),' Distr. General, 19 April 1993.

IAEA, S/26685, 'Note by the SG,' 3 November 1993, annexed to it a 'Letter dated 29 October 1993 from the Acting Director General of the IAEA addressed to the SG,' enclosed with it the 'Fourth Report of the DG of the IAEA on the Implementation on the Agency's Plan for Future Ongoing Monitoring and Verification of Iraq's Compliance with Paragraph 12 of Resolution 687 (1991),' Distr. General, 3 Novemeber 1993.

IAEA S/1995/844, DG Report, 'Eighth Report of the DG of the IAEA on the Implementation of the Agency's Plan for Future Ongoing Monitoring and Verification of Iraq's Compliance with Paragraph 12 of Resolution 687 (1991),' Distr. General, 6 October 1995.

IAEA, GOV/2003/40, DG Report on the 'Implementation of the NPT Safeguards Agreement in the Islamic Republic of Iran,' submitted to the Board of Governors of the IAEA on 6 June 2003 and derestricted on 19 June 2003.

IAEA, GOV/2003/63, DG Report on the 'Implementation of the NPT Safeguards

Agreement in the Islamic Republic of Iran,' submitted to the Board of Governors on 26 August 2003 and derestricted on 12 September 2003.
IAEA, GOV/2003/75, DG Report, 'Implementation of the NPT Safeguards Agreement in the Islamic Republic of Iran,' submitted to the Board of Governors of the IAEA on 10 November 2003 and derestricted on 26 November 2003.
IAEA, GOV/2004/11, DG Report, 'Implementation of the NPT Safeguards Agreement in the Islamic Republic of Iran,' submitted to the Board of Governors of the IAEA on 24 February 2004 and derestricted on 13 March 2004.
IAEA, GOV/2004/34, DG Report to the Board of Governors, 'Implementation of the NPT Safeguards Agreement in the Islamic Republic of Iran,' 1 June 2004 and derestricted on 18 June 2004.
IAEA, GOV/2004/60, DG Report, 'Implementation of the NPT Safeguards Agreement in the Islamic Republic of Iran,' submitted to the Board of Governors of the IAEA on 1 September 2004, and derestricted on 18 September 2004.
IAEA, GOV/2004/83, Report by the DG on the 'Implementation of the NPT Safeguards Agreement in the Islamic Republic of Iran,' submitted to the Board of Governors on 15 November 2004 and derestricted 15 November 2004.
IAEA, GOV/2005/64, Report by the DG on the 'Implementation of the NPT Safeguards Agreement in the Islamic Republic of Iran and Related Board Resolutions,' 11 August 2005 and derestricted on 11 August 2005.
IAEA, GOV/2005/67, Report by the DG, on the 'Implementation of the NPT Safeguards Agreement in the Islamic Republic of Iran,' 2 September 2005 and derestricted on 24 September 2005.
IAEA, GOV/2005/77, Report by the DG on the 'Implementation of the NPT Safeguards Agreement in the Islamic Republic of Iran,' 24 September 2005 and derestricted on 24 September 2005.
IAEA, GOV/2006/15, Report by the DG on the 'Implementation of the NPT Safeguards Agreement in the Islamic Republic of Iran,' 27 February 2006 and derestricted on 8 March 2006.
IAEA, GOV/2006/27, Report by the DG on the 'Implementation of the NPT Safeguards Agreement in the Islamic Republic of Iran,' 28 April 2006 and derestricted on 12 June 2006.
IAEA, GOV/2006/53, Report by the DG on the 'Implementation of the NPT Safeguards Agreement in the Islamic Republic of Iran,' on 31 August 2006 and derestricted on 14 September 2006.
IAEA, GOV/2006/64, Report by the DG on the 'Implementation of the NPT Safeguards Agreement in the Islamic Republic of Iran,' on 14 November 2006 and derestricted on 23 November 2006.
IAEA, GOV/2007/7, Report by the DG on the 'Cooperation between the Islamic Republic of Iran and the Agency in the light of the UN SCR 1737 (2006),' 9 February 2007 and derestricted on 8 March 2007.
IAEA, GOV/2007/8, Report by the DG on the 'Implementation of the NPT Safeguards Agreement and Relevant Provisions of SCR 1737 (2006) in the Islamic Republic of Iran,' 22 February 2007 and derestricted on 7 March 2007.
IAEA, GOV/2007/22, Report by the DG on the 'Implementation of the NPT Safeguards Agreement and Relevant Provisions of SC Resolutions in the Islamic Republic of Iran,' on 23 May 2007.

IAEA, GOV/2007/48, Report by the DG on the 'Implementation of the NPT Safeguards Agreement in the Islamic Republic of Iran,' 30 August 2007 and derestricted on 12 September 2007.

IAEA, GOV/2007/58, Report by the DG on the 'Implementation of the NPT Safeguards Agreement and Relevant Provisions of Security Council Resolutions 1737 (2006) and 1747 (2007) in the Islamic Republic of Iran,' 15 November 2007 and derestricted on 23 November 2007.

IAEA, GOV/2008/4, Report by the DG on the 'Implementation of the NPT Safeguards Agreement and Relevant Provisions of Security Council Resolutions 1737 (2006) and 1747 (2007) in the Islamic Republic of Iran,' 22 February 2008 and derestricted on 5 March 2008.

IAEA, GOV/2008/15, Report by the DG on the 'Implementation of the NPT Safeguards Agreement and Relevant Provisions of Security Council Resolutions 1737 (2006), 1747 (2007) and 1803 (2008) in the Islamic Republic of Iran,' 26 May 2008.

IAEA, GC (45)/INF/5, General Conference, 'Nuclear Technology Review 2001,' Forty-Fifth Regular Session Sub- Item 17(d) of the Provisional Agenda (GC (45)/1), August 2001, cited from Ann Bisconti, 'Overcoming the Perception Gap: Communicating Public Support for Nuclear Energy,' presentation at the Americas Nuclear Energy Symposium, 11–13 December 2000, Miami, FL; NucNet News No. 444—A, 21 Dec. 2000. REUTERS Business Briefing, 1-3-2001, "70% of Germans support use of nuclear power—study," Sources: PLATT'S COMMODITY NEWS 02/01/2001; REUTERS Business Briefing, 1-2-2001, "Koreans concerned about safety of nuke plants—survey," Sources: THE KOREA HERALD 28/12/2000; NucNet Background No. 33, 5 Jan. 2001.

IAEA, 'Statement by H.E. Reza Aghazadeh, Vice-President of the Islamic Republic of Iran and the President of the Atomic Energy Organisation of Iran at the 46th General Conference of the IAEA,' Vienna, 16 September 2002.

ICJ, Reports of Judgments, Advisory Opinions, Orders 'Reparation of Injuries Suffered in the Service of the UN,' Advisory Opinion of 11 April 1949.

ICJ, Reports of Judgements, Advisory Opinions, Orders, 'Legal Consequences for States of the Continued Presence of South Africa in Namibia (South-West Africa), Notwithstanding Security Council Resolution 276 (1970),' Advisory Opinion of 21 June 1970.

ICJ, Reports of Judgements, Advisory Opinions, Orders, 'Interpretation of the Agreement of 25 March 1951 between the WHO and Egypt,' Advisory Opinion of 20 December 1980.

Ijaz, Mansoor, and R. James Woolsey, 'Cut Supply Lines that Fuel Pyongyang's Nuclear Dreams,' Los Angeles Times, 12 January 2003.

Isard, W., 'Some Economic Implications of Atomic Energy,' Quarterly Journal of Economics, Volume 62, MIT Press, Cambridge, February 1948.

Isard, W. and J. Lansing, 'Comparisons of Power Cost for Atomic and Conventional Steam Stations,' Review of Economics and Statistics, Volume 31, MIT Press, Cambridge, 1949.

Isard, W. and V. Whitney, 'Atomic Power: An Economic and Social Analysis,' Blakiston Press, New York, US, 1952.

Karayiannopoulos, Fotis A., Barrister and Member of the Board of Appeals of the

Community Plant Variety Office, 'International Environmental Law,' Greece, http://users.hol.gr/~bio/HTML/PUBS/budapest/Fotis.

Karsh, Efraim and Martin Navias, 'Israeli Nuclear Weapons and Middle East Peace,' in Efraim Karsh (ed.), 'Between War and Peace: Dilemmas of Israeli Security,' Frank Cass & Co. Ltd., London, 1996.

Khadduri, Imad, 'Iraq's Nuclear Non-Capability,' YellowTimes.Org, 21 November 2002. www.yellowtimes.org/article.php?sid=874.

Khan, R., 'The Legal Status of the Resolutions of the UN General Assembly,' The Indian Journal of International Law, Volume 19, Indian Society of International Law, New Delhi, 1979.

Klare, Micheal T., 'Controlling the Trade in Conventional Weapons,' Transnational Law & Contemporary Problems, Volume 2, A Journal of University of Iowa College of Law, Iowa, Fall 1992.

Kokoski, Richard, 'Technology and the Proliferation of Nuclear Weapons,' Stockholm International Peace Research Institute, Oxford University Press, New York, 1995.

Koplow, David A., 'The Jurisprudence of Non-Proliferation: Taking International Law Seriously,' Transnational Law & Contemporary Problems, Volume 2, A Journal of University of Iowa College of Law, Iowa, Fall 1992.

Krosney, Herbert, 'Deadly Business—Legal Deals and Outlaw Weapons: The Arming of Iran and Iraq, 1975 to the Present,' Four Walls Eight Windows, New York, 1993.

Leonard, James F., 'Strengthening the Treaty on the Non-Proliferation of Nuclear Weapons,' Transnational Law and Contemporary Problems—A Journal of the University of Iowa, Volume 2, Number 1, Spring 1992.

Marom, Ran, 'Israel's Position on Non-Proliferation,' the Leonard Davis Institute for International Relations, University of Jerusalem, June 1996.

Mazer, Brian M., 'The Evolution of Canadian Agreements for Cooperation in Nuclear Weapons and Technology,' The Canadian Yearbook of International Law, Volume XIII, The University of British Columbia Press, Vancouver, B.C, 1980.

Meyrowitz, Elliott L., 'The Laws of War and Nuclear Weapons,' from Arthur Selwyn Miller and Martin Feinrider, 'Nuclear Weapons and Law,' Greenwood Press, 1984.

Monahan, Bill, 'Giving the Non-Proliferation Teeth: Strengthening the Special Inspection Procedures of the IAEA,' Virginia Journal of International Law, Volume 33, Virginia, 1992–1993.

Miller, Arthur Selwyn and Martin Feinrider, 'The Laws of War and Nuclear Weapons,' Contributions in legal studies, no. 31. Greenwood Press, Westport, Conn., 1984.

Milhollin, Gary and Valerie Lincy, 'Iran's Nuclear Card,' Commentary Magazine, New York, February 2004, http://www.wisconsinproject.org/pubs/articles/2004/Commentary.pdf

Mullenbach, P. 'Civilian Nuclear Power: Economics Issues and Policy Formation,' Twentieth Century Fund, New York, 1963.

Muller, Harold, 'Europe's Leaky Borders,' Bulletin of the Atomic Scientists, Volume 49, No. 5, Chicago, June 1993. http://www.the bulletin.org/index.html.

Muntzing, L. Manning, 'Philosophical Goals of Nuclear Regulation,' Nuclear Law Bulletin, Bulletin 17, Nuclear Energy Agency, OECD, Paris, April 1976.

Nanda, Ved P. and William M. Evans, 'Nuclear Proliferation and the Legality of Nuclear Weapons,' University Press of America, Inc., Maryland, 1995.

Nichols, Jere, Chemical Engineer, Oak Ridge National Laboratory (retired), 'Understanding the Lessons of Nuclear Inspections and Monitoring in Iraq: A Ten-Year Review,' A Seminar sponsored by the Institute for Science and International Security, 'Uncovering the Secret Program—Initial Inspections,' 14–15 June 2001, Transcript date: 4 September 2001.http://www.isis-online.org/publications/iraq/nichols.html.

Nolan, Janne, 'Technology and Non-Proliferation in a Changing World Order,' Transnational Law and Contemporary Problems, Volume 2, Number 1, A Journal of The University Iowa College of Law, Iowa, 1992.

Norbert Pelzer, 'The Nature and Scope of International Cooperation in Connection with the Peaceful Uses of Nuclear Energy, and Its Limits—An Assessment,' Nuclear Law Bulletin, Volume 27, Nuclear Energy Agency, OECD, Paris, June 1981.

NPT Review and Extension Conference, 'Final Document of the 1995 Review and Extension Conference of the Parties to the NPT,' New York, 1995.

NPT Review Conference, 'Final Document of the 2000 Review Conference of the Parties of the NPT,' New York, 2000.

Nuclear Energy Agency, 'Nuclear Legislation, Analytical Study, Regulation and Framework for Nuclear Activities,' Nuclear Law Bulletin, OECD, Volume II, Paris, 1984.

Nuclear Energy Agency, 'The Regulation of Nuclear Trade: Non-Proliferation Supply Safety: International Aspects,' Volume 1, OECD, Paris, 1988.

Nuclear Energy Agency, 'The Regulation of Nuclear Trade: Non-Proliferation Supply-Safety: National Regulations,' Volume 2, OECD, Paris, 1988.

Nuclear Energy Policy Study Group, and Spurgeon M. Keeny, 'Nuclear Power Issues and Choices: Report of the Energy Policy Study Group,' sponsored by the Ford Foundation, Ballinger, Cambridge, Mass, 1977.

Nuclear Issues Briefing, 'World Energy Needs and Nuclear Power,' Paper 11, July 2002. http://www.uic.com.au/nip11.html

Nuclear Issues Briefing, 'Nuclear Power in the World Today,' Paper 7, June 2003. http://www.uic.com.au/nip07.htm

Nuclear Suppliers Group, 'International Seminar on the Role of Export Controls in Nuclear Non-Proliferation,' Vienna International Centre, 7–8 October 1997.

OECD homepage, http://www.nea.fr/html/nea/flyeren.html.

Prawitz, Jan, 'Nuclear-Weapon-Free Zones: Their Added Value in a Strengthened International Safeguards System,' in E. Häckel and G. Stein, (eds.), 'Tightening the Reins: Towards a Strengthened International Nuclear Safeguards System,' Berlin, 2000.

Pfaffelhuber, Josef K., Ministerial RAT, Federal Ministry of Education and Science, Germany, 'New Trends in Atomic Law,' Nuclear Law Bulletin, No. 10, Nuclear Energy Agency, OECD, Paris, 1972.

Plant, Sir Arnold, 'The Economics of Nuclear Energy,' in Otto Klineberg (ed.), 'Social implications of the Peaceful Uses of nuclear Energy,' UNECSO, Paris, 1964.

Poong-Eil Juhn, Juergen Kuptiz, and John Cleveland, 'Advanced nuclear power plants: Highlights of global development,' IAEA Bulletin, Volume 39, Number 2, Vienna, 1997. http://www.iaea.org/worldatom/Periodicals/Bulletin/Bull392/author.

Perricos, Dimitri, Former Deputy Leader of the IAEA Action Team, Director of the Division of Planning and Operations at UNMOVIC in New York, 'Understanding the Lessons of Nuclear Inspections and Monitoring in Iraq: A Ten-Year Review,' A Seminar sponsored by the Institute for Science and International Security, 'Uncovering the Secret Program—Initial Inspections,' 14–15 June 2001, http://www.isis-online.org/publications/iraq/perricos.

Pilat, Josef, 'The Major Suppliers: A Baseline for Comparison,' in William C. Potter (ed.), 'International Trade and Non-Proliferation: The Challenge of the Emerging Suppliers,' Monetary Institute of International Studies, Published under the auspices of the Centre for The International and Strategic Affairs, University of California, Lexington Books, D.C. Heath and Company, Lexington, Massachusetts/Toronto, 1990

Potter, William C., 'International Trade and Non-proliferation—the Challenge of the Emerging Suppliers,' Monetary Institute of International Studies, Published under the auspices of the Centre for The International and Strategic Affairs, University of California, Lexington Books, D.C. Heath and Company, Lexington, Massachusetts/Toronto, 1990.

Priddle Robert, 'Energy and Sustainable Development,' IAEA Bulletin 4, Vienna, 1 January 1999.

Rabl Ari and Mona Dreicer, 'An Eye to the Future—Long-Term Health and Environmental Impacts on Energy Systems,' IAEA Bulletin, Bulletin 4, Vienna, 1 January 1999.

Raghavan, C.G., 'The United Nations Charter and the Use of Force in Inter-state Relations,' in R.C. Hingorani (ed.), 'International Law through United Nations—Seminar on the Role of the United Nations in the Development of International Law Patna University 1970,' N.M. Tripathi Private Limited, Bombay, 1972.

Tarik Raouf, Section Head, Verification and Security Policy Coordination, Office of External Relations and Policy Coordination at the IAEA. This statement was given in an interview concluded at IAEA Headquarters, 2004. (Meeting)

Rauschning, Dietrich, Katja Wiesbrock and Martin Lailach, 'Key resolutions of the United Nations General Assembly, 1946–1996,' Cambridge University Press, Cambridge, 1997.

Renshaw, E., 'Atomic Power: Research Costs and Social Returns,' Land Economics, Volume 35, University of Wisconsin Press, Wis., 1959.

Repertoire of the practice of the SC, Chapter V: Subsidiary Organs of the SC, volumes covering years from 1946 until 2003, http://www.un.org/Depts/dpa/repertoire/index.html.

Rhinelander, John B. and Adam M. Scheinman, 'At the Nuclear Crossroads, Choices about Nuclear Weapons and Extension of the Non-Proliferation

Treaty,' The Lawyers Alliance for World Security, University Press of America, Inc, Maryland, 1995.

Rockwood, Laura, 'Legal Instruments Related to the Application of the Safeguards,' OPANAL, Jamaica, 1998–2000, http://www.opanal.org/Articles/Jamaica/jam-Rockwood.htm.

Rometsch, R., 'International Safeguards on the Peaceful Uses of Nuclear Material,' Nuclear Law Bulletin, No. 13, Nuclear Energy Agency, OECD, Paris, 1974.

Rosen, Mark E., Commander, JAGC, US Navy, 'Nuclear-Weapon-Free Zones,' Naval War College Review, Autumn 1996, Volume LXIX, No. 4, http://www.nwc.navy.mil/press/Review/1996/autumn/free-a96.html.

Roshandel, Jalil, 'Iran, Nuclear Technology and International Security,' The Iranian Journal of International Affairs, Volume 8, Issue 1, 1996.

Rydell, Randy J., 'Studying the Emerging Nuclear Suppliers: A Framework for Analysis,' in William C. Potter (ed.), 'International Nuclear Trade and Non-Proliferation—The Challenge of The Emerging Suppliers,' Monetary Institute of International Studies, Published under the auspices of the Centre for The International and Strategic Affairs, University of California, Lexington Books, D.C. Heath and Company, Lexington, Massachusetts/Toronto, 1990.

Sanders, Benjamin and Ha Vinh Phuong, 'International Safeguards,' Nuclear Law Bulletin, Volume 18, Nuclear Energy Agency, OECD, Paris, 1976.

Sanders, Benjamin, Founder and Executive Chairman of the Programme for Promoting Nuclear Non-Proliferation. Former official, of Netherlands Ministry of Foreign Affairs; IAEA Safeguards Department, and United Nation's Department for Disarmament Affairs. This statement was given in an interview held at UN in 2004. (Meeting)

Sands, Amy, 'Emerging Nuclear Suppliers: What's The Beef?' in William C. Potter (ed.), Potter, William, 'International Trade and Non-Proliferation: The Challenge of the Emerging Suppliers,' Monetary Institute of International Studies, Published under the auspices of the Centre for The International and Strategic Affairs, University of California, Lexington Books, D.C. Heath and Company, Lexington, Massachusetts/Toronto, 1990.

Schiff, Ze'ev, 'Weapons of Mass Destruction and the Middle East, View from Israel,' James A. Baker III, Institute for Public Policy, Rice University, Princeton Architectural Press, New York, March 2003.

Schachter, Oscar, 'Peaceful Uses of Nuclear Energy: Environmental, Security and Safety Considerations,' Columbia Journal of Transnational Law, Volume 16, Columbia Society of International Law, New York, 1977.

Schachter, Oscar, 'United Nations Law,' American Journal of International Law, Volume 88, the American Society of International Law, Washington D.C., 1994.

Scheinman, Lawrence, 'The Role of Multilateral Regimes in Non-Proliferation,' Transnational Law and Contemporary Problems, Volume 2, Number 1, A Journal of the University Iowa, Iowa, 1992.

Schmidt, Fritz, 'Zangger Committee,' paper presented before the 2nd Seminar on Export Controls in Nuclear Non-Proliferation held at UN in New York, 8 to 9 April 1999.

Schwebel, S.M., 'The Effect of Resolutions of the General Assembly on Customary International Law,' the American Society of International Law, Washington, D.C., 1979.

SC, Resolution 699 (1991), adopted by the SC in its 2994th Meeting, on 17 June 1991. Distr. General, S/Res/699 (1991).

SC, Resolution 707 (1991) adopted by the SC at its 3004th Meeting, on 15 August 1991. Distr. General, S/Res/707 (1991).

SC, Resolution 1115 (1997) adopted by the SC at its 3792nd Meeting, on 21 June 1997. Distr. General, S/Res/1115 (1997).

SC, Resolution 1051 (1996) adopted by the SC at its 3644th Meeting, on 27 March 1996. Distr. General S/RES/1051 (1996).

SC, Resolution 1137 (1997) adopted by the SC at its 3381st Meeting on 12 November 1997. Distribution General, S/Res/1137 (1997).

SC, S/22837, 'Note by the UN SG,' dated 25 July 1991, annexed to it a 'Letter dated 25 july 1991 from the DG of the IAEA addressed to the SG,' enclosed with it the 'Third IAEA on-Site Inspection Report in Iraq (7–18 July 1991) under SC Resolution 687 (1991),' 25 July 1991.

SC, S/22986, 'Note by the UN SG,' dated 28 August 1991 to the SC, annexed to it a 'Letter dated 27 August 1991 from the DG of the IAEA addressed to the SG,' enclosed with it the 'Report on the Fourth IAEA on-Site Inspection in Iraq (27 July–10 August 1991) under Security Council Resolution 687 (1991),' 28 August 1991.

SC, S/23122, 'Note by the UN SG,' dated 8 October 1991, annexed to it a 'Letter dated 3 October 1991 from the DG of the IAEA addressed to the SG,' enclosed with it the 'First Report on the Sixth IAEA on-Site Inspection in Iraq (22–30 September 1991) under SC Resolution 687 (1991),' 11 December 1991.

SC, S/23283, 'Note by the UN SG,' dated 11 December 1991, annexed to it a 'Letter dated from the DG of the IAEA addressed to the SG,' enclosed with it the 'Report on the Eighth IAEA on-Site Inspection in Iraq (11–18 November 1991) under the SC Resolution 687 (1991),' 11 December 1991. http://www.iaea.org/worldatom/Programmes/ActionTeam/chronology.html.

SC, S/23505, 'Note by the SG to the SC,' dated 30 January 1992, annexed to it a 'Letter dated 28 January 1992 from the DG of the IAEA addressed to the SG,' enclosed with it the 'Report on the Ninth IAEA on-Site Inspection in Iraq (11–14 January 1992) under SC Resolution 687 (1991),' 30 January 1992.

SC, S/23644, 'Note by the UN SG,' 26 February 1992, annexed to it a 'Letter dated 25 February 1992 from the DG of the IAEA addressed to the SG,' enclosed with it the 'Report on the Tenth IAEA on-Site Inspection in Iraq (5–13 February 1992) under SC Resolution 687 (1991) SC,' 26 February 1992.

SC, S/23947, 'Note by UN SG,' dated 22 May 1992, annexed to it a 'Letter dated 12 May 1992 from the Acting DG of the IAEA addressed to the SG,' enclosed with it the 'Report on the Eleventh IAEA on-Site Inspection in Iraq (7–15 April 1992) under SC Resolution 687 (1991),' 22 May 1992.

SC, S/24981, 'Note by UN SG,' dated 17 December 1992, annexed to it a 'Letter dated 10 December 1992 from the Acting DG of the IAEA addressed to the SG,' enclosed with it the 'Report on the Fifteenth IAEA on-Site Inspection in Iraq (8–18 November 1992) under SC Resolution 687 (1991),' 17 December 1992.

SC, S/1994/31, 'Note by the UN SG,' dated 14 January 1994, annexed to it a 'Letter dated 11 January 1994 from the DG of the IAEA,' enclosed with it the 'Report on the Twenty-Second IAEA on-Site Inspection in Iraq under Security Council Resolution 687 (1991),' 14 January 1994.

SC, S/1994/355, 'Note by the UN SG,' dated 25 March 1994, annexed to it a 'Letter dated 23 March 1994 from the Acting DG of the IAEA,' enclosed with it the 'Report Covering the Period 4–11 February 1994 on the Twenty-Third IAEA on-Site Inspection in Iraq under Security Council Resolution 687 (1991),' 25 March 1994.

SC, S/1994/650, 'Note by the UN SG,' dated 1 June 1994, annexed to it a 'Letter dated 30 May from the DG of the IAEA,' enclosed with it the 'Report on the Twenty-Fourth IAEA on-Site Inspection in Iraq under Security Council Resolution 687 (1991),' 1 June 1994.

SC, S/1994/1001, 'Note by the UN SG,' dated 26 August 1994, annexed to it a 'Letter dated 23 August 1994 from the Acting DG General of the IAEA,' enclosed with it the 'Report on the Twenty-Fifth IAEA on-Site Inspection in Iraq under Security Council Resolution 687 (1991),' 26 August 1994.

SC, S/1994/1206, 'Note by the UN SG,' dated 22 October 1994, annexed to it a 'Letter Dated 18 October 1994 from the Acting DG of the IAEA,' enclosed with it the 'Report on the Twenty-Sixth IAEA on-Site Inspection in Iraq under Security Council resolution 687 (1991),' 22 October 1994.

SC, S/1994/144, 'Note by the UN SG,' dated 22 December 1994 annexed to it a 'Letter dated 19 December 1994 from the Acting DG of the IAEA,' enclosed with it the 'Report on the Twenty-Seventh IAEA on-Site Inspection in Iraq under Security Council Resolution 687 (1991),' 22 December 1994.

SC, S/1995/1003, 'Note by the UN SG,' dated 1 December 1995, annexed to it a 'Letter dated 24 November 1995 from the DG of the IAEA,' enclosed with it the 'Report on the Twenty-Eighth IAEA on-Site Inspection in Iraq under Security Council Resolution 687 (1991),' 1 December 1995.

SC, S/1996/14, 'Note by the UN SG,' dated 10 January 1996, annexed to it a 'Letter dated 10 January 1996 from the DG of the IAEA,' enclosed with it the 'Report on the Twenty-Ninth IAEA on-Site Inspection in Iraq under Security Council Resolution 687 (1991),' 10 January 1996.

SC, S/1997/779, 'Note by the UN SG,' dated 8 October 1997, annexed to it a 'Letter dated 6 October 1997 from the DG of the IAEA,' enclosed with it the 'Fourth Consolidated Report of the DG of the IAEA under Paragraph 16 of the SC Resolution 1051 (1996),' 8 October 1997.

SCR, S/RES/1762 (2007) adopted by the SC in its 5710th Meeting, on 29 June 2007.

SC, S/2005/652, 'Letter dated 14 October 2005 from the SG addressed to the President of the SC,' dated 17 October 2005, attached to it the 'Letter dated 13 October 2005 from the DG of the IAEA addressed to the President of the SC,' 17 October 2005.

SC, S/PRST/2006/15, 'Statement by the President of the SC,' 29 March 2006.

SC, S/2006/521, dated 13 July 2006, "Letter dated 13 July 2006 from the Permanent Representative of France to the UN addressed to the President of the SC."

SC, S/RES/1696 adopted by the SC in its 5500th Meeting, dated 31 July 2006.

SC, S/RES/1737 dated 27 December 2006, adopted by the SC at its 5612th Meeting, 23 December 2006.

SC, S/RES/1747 adopted by the Security Council at its 5647th Meeting, 24 March 2007.

SC, S/RES/1803, adopted by the Security Council at its 5848th Meeting, 3 March 2008.

Sethna, J. 'International Legal Controls and Sanctions Concerning the Production and Use of Atomic Energy,' Bombay, 1966.

Seymour, Hersh, 'The Samson Option—Israel's Nuclear Arsenal and the American Foreign Policy,' Random House, New York, 1991.

Shaikh, Farzana, 'Pakistan's Nuclear Bomb: Beyond the Non-Proliferation Regime,' International Affairs, the Royal Institute of International Affairs, Volume 78, Number 1, Blackwell Publishers Limited, Malden, January 2002.

Shaker, Mohamed I., 'The Nuclear Non-Proliferation Treaty: Origin and Implementation 1959–1979, the Treaty on the Non-Proliferation of Nuclear Weapons,' Oceana Publications, London, 1980.

Shaker, Mohamed I., 'The Middle East Issue; Possibilities of a Nuclear-Weapons-Free Zone,' published in 'XXX Anniversary of the Opening for Signature of the Treaty of Tlatelolco for the Prohibition of Nuclear Weapons in Latin America and the Caribbean—The Treaty of Tlatelolco and the Nuclear-Weapon-Free Zones: An Example to follow,' 14 February 1997, http://www.opanal.org/Articles/Articulos-i.htm.

Sloan, Blaine, 'General Assembly Resolutions Revisited (Forty Years Later),' The British year book of international law, Volume 58, Oxford University Press, 1987.

Sloan, Blaine, 'UN GA Resolutions in Our Changing World,' Transnational Publishers, Inc.—Ardsley-on-Hudson, New York, 1991.

Sloss, David, 'It's Not Broken, So Don't Fix It: The International Atomic Energy Agency Safeguards System and the Nuclear Non-Proliferation Treaty,' Virginia Journal of International Law, Volume 35, Issue 4, Virginia, 1995.

Smith, Jacqueline R., and Lewis A. Dunn, 'Overview: The Main Proliferation Risks,' Transnational Law & Contemporary Problems, Volume 2, A Journal of University of Iowa College of Law, Iowa, Fall 1992

Sonnenfeld, Renata, 'Resolutions of the United Nations Security Council,' Martinus Nijhoff Publishers, Dordrecht, 1988.

Steinberg, Gerald M., 'Starting over: The Prospects for Regional Security and Arms Control in the Middle East in the Next Decade,' UNIDIR, Disarmament Forum, Issue 2, Geneva, 2001.

Stratford, Richard J.K., 'The Practice of Export Controls: Effects on Trade and Peaceful Nuclear Activities,' paper presented before the 2nd Seminar on Export Controls in Nuclear Non-Proliferation held at UN in New York, 8–9 April 1999.

Sugiyama, Shinsuke J., 'Peaceful Uses of Nuclear Energy and the New Era for Japan-US Cooperation—History and Major Issues of the New Japan-US Agreement,' the Japanese Annual of International Law, No. 31, Tokyo, 1988.

Timmerman, Kenneth R., Director, the Middle East Data Project, 'Iran's Nuclear Program: Myth and Reality,' Middle East Data Project, Inc., 1995. A paper presented before The Italian Union of Scientists for Disarmament (USPID)

Sixth International Castiglioncello Conference, titled 'Fifty Years after Hiroshima,' Castiglioncello, Italy, 30 September 1995.

Thomson, David B., 'A Guide to the Nuclear Arms Control Treaty,' Los Alamos Historical Society, Los Alamos, New Mexico, 2001.

UN, 'Assuring the Success of the Non-Proliferation Treaty Extension Conference,' Excerpts from the panel discussions organised by the NGO Committee on Disarmament, Inc., at the Conference held at the United Nations in New York, 20–21 April 1994, United Nations Publications, Sales No. E.94.IX.9.

United Nations Department for Disarmament Affairs Press Kit, 'Review Conference of the Parties to the Treaty on the Non-Proliferation of Nuclear Weapons—fact sheet 2,' 24 April–19 May 2000, New York, DPI/2085—February 2000–2,500.

US Policy in Atomic Energy Development, 'Atoms for Power,' the American Assembly, Columbia University, New York, December 1957.

Niksch, Larry A., 'North Korea's Nuclear Weapons Program,' State Department, CRS Issue Brief for Congress, updated 5 October 2006, http://fpc.state.gov/documents/organization/74904.pdf.

Van den Borre, Tom and Roland Carchon, 'Preventing the Proliferation of Nuclaer Weapons: 50 Years of Atoms for Peace,' Nuclear Law Bulletin No. 57, Nuclear Energy Agency, OECD, Paris, June 1996.

Weinberg, A. 'How Shall We Establish a Nuclear Power Industry in the United States?' Bulletin of Atomic Scientists, Volume 9, Symposium on 'Atomic Energy and Private Enterprise,' Chicago, May 1953.

Weiss, Leonard, 'Pakistan: it's déjà vu all over again,' Bulletin of Atomic Scientists, Volume 60, No. 3, Chicago, May/June 2004.

Wood, Michael C., 'The Interpretation of Security Council Resolutions,' Max Planck Yearbook of United Nations Law, Volume 2, Kluwer Law International, London, 1998.

Yager, Joseph A. and Eleanor Steinberg, 'Energy and US foreign policy—A Report of the Energy Policy Project of the Ford Foundation,' Ballinger Publishing Company, Cambridge, Mass., 1974.

Yaniv, Avner, 'Non Conventional Weapons and the Future of Arab-Israeli Deterrence,' in Efraim Karsh (ed.), 'Between War and Peace: Dilemmas of Israeli Security,' Frank Cass & Co. Ltd., London, 1996.

Young, Warren 'Atomic Energy Costing,' Kluwer Academic Publishers, 1998, quoted from a document sent by Roy Harrod to Lord Cherwell Papers for the latter comments on 29 May 1947 [Harrod to Cherwell, Cherwell Papers j 71/16, j71/15

INDEX